BRINGING IN THE WOOD

The way it was at Chesapeake Corporation

by Mary Wakefield Buxton

Library of Congress Catalog Number
99-76201

International Standard Book Number
1-880902-13-3

Manufactured in the United States of America
in association with
Rappahannock Press Inc., 276 Virginia Street,
Urbanna, Virginia 23175-0549

First Edition

Dedication:

This book is dedicated to the Olsson family, with love and appreciation for all that they gave to the Company, employees, the town of West Point, and the state of Virginia.

"Never forget, my son, our most important asset is our people."
Advice from CEO Elis Olsson to the new CEO Sture Olsson, supervisors meeting, West Point, Va., 1946

Contents:

Notes from the Author:

From the very beginning of this book project I encouraged and Chesapeake supported the idea that this story not be some politically correct "Company line" or a dull recount of a zillion boring facts. Rather, we wanted actual stories told by the employees themselves of what life was like bringing in the wood based on their individual memories, thoughts and feelings.

Thus, we have a special book written by the people themselves of their accounts as to the way things were. And like every book that tells individual stories, these stories do not always mesh perfectly. People see things differently. This individual interpretation of how things were is what we rejoice in and celebrate in publishing this book.

Those who are looking for exact factual history, an "official" Company handbook, some particular agenda or even some interpretation of some special interest have come to the wrong book. We created this book with only one purpose in mind: To relate the individual at Chesapeake and to tell his stories as to the way things were. And in so doing, we hold dear our founder's highest asset: the individual employee and his special contribution to the Company.

Mary Wakefield Buxton
Urbanna, Virginia
August 31, 1999

To avoid any confusion regarding the Company names used in this book, please note the following:

The original corporate name for the company Elis Olsson founded in West Point, Virginia, in 1918 was "The Chesapeake Corporation of Virginia." In 1980, however, the name of the Company was changed to the present "Chesapeake Corporation" in order to reflect a new global status.

The name of the original department within the Company once known as the "Woodlands Division" was also changed in the early 1980s to "Chesapeake Forest Products Company."

Acknowledgments

This book was the brainchild of Jack King and Jim Vadas of the Woodlands Division of Chesapeake Corporation who together for many years considered the idea of having a book written on the history of Woodlands. Without the dedication of these two men, this book would never have been published.

Two other foresters, Sharon Miller and Dick Brake, both retired from Chesapeake Corporation, also served on the "book committee" and gave untold hours of service to see that every detail was correct. Also, Ida Dawson and Thelma Downey of the Woodlands staff helped set up interviews and assisted in many ways to see this book project completed.

I would also like to thank each person who gave me an interview or provided me information for all the wonderful material that went into this history. Lastly, I am appreciative of Chesapeake Corporation who generously funded this project in the name of its past and present Woodlands' employees and friends.

MWB

Introduction

Ever since the beginning of time and the first appearance of an animal that resembled modern man, he has been bringing in the wood.

In the beginning, long before 20th century foresters emerged in present civilization armed with every sort of know-how regarding the art of growing and harvesting trees, man has needed wood. Indeed, wood became a major source for his very survival, livelihood and eventual emergence as "king of the beasts."

It is easy to imagine early man standing in primeval forest and smelling the pungent aroma for the first time of the distant smoke of a forest fire. One can almost see him standing alert, twitching his nose and turning in the direction of the smoke. Surely, as he curiously approached the roar and crackle of fire, he noticed the instant destruction as the red gold flame engulfed the trees and surrounding brush. Certainly he felt the intense heat and considered warmth for his own cave.

It followed soon after that early man picked up a burning ember and returned to his cave. From there he added the bark and twigs to make his very first campfire. This fire provided the first heat for his home along with energy to cook his food or boil his water. Before long, early man was going into the woods each day in the simple act of bringing in the wood.

Later, early man learned that wood had other magical uses. He could snap off certain limbs and crudely affix them together, spread animal skin or bark or leaves or mud over them and build a rough hovel or tent. Wood first allowed man to exit his cave. He learned a stick could be broken in such a way to be made into a rough tool, perhaps the end even sharpened in order to stab a fish or some small animal. Next, man used wood in order to fashion a club or a spear in order to attack other tribes or to defend himself.

Not long after that, man began using wood for primitive furniture, very probably starting with large logs dragged into the cave or close to the campfire to act as stools or benches. Later, man discovered that wood floated on water and a log could carry a man across a river. From that point on, the first primitive water craft was created, perhaps a raft or dugout canoe, and this allowed man his very first opportunity to escape land and journey forth across the sea.

Man learned his lesson quickly, a lesson that is still true today. Wood was his most important staple. Each century brought on new uses and greater need for wood. Even in today's hi-tech time, man cannot live without wood. Thus man still goes into the woods each day and brings back the wood. This act, as simple as it is, is a very important link to our modest beginnings. In spite of rapidly changing global societies, we still need our daily wood.

So why do we tell this story? Because bringing in the wood is a task that is just as essential today as it was in primitive times. Bring-ing in the wood links us to our very beginnings; indeed, connects us to our roots. More than that, it reminds us once again of who we are and how we sprang to our modern day exalted position on earth. As this realization triggers great awe, it also reminds us of our humble past and our great responsibility to care for this earth and the treasure of trees that have built our civilization.

This book takes us to the center of all life, directly into the woods. There we stop for a moment from our busy lives and give thanks for the trees that have given so much to our continuing existence.

Nowhere are we closer to Mother Earth than when we are in the woods and amongst the trees. We stop to listen to the chatter of squirrel from the limb and the sweet song of bird on the vine. We look up and see the rich tapestry of great blue sky laced with green leaves. And we are filled with a great sense of appreciation for trees and all the blessings they have delivered to this world.

This book relates how bringing in the wood was done in this century in one small part of the world and from one small paper company that started in West Point, Virginia, in the early 1900s. We tell of the close friendships within the Company, the highly cohesive team of men and women, and the hard work and sacrifice that went into the completion of their daily tasks. We tell how this fundamental need provided jobs, money and livelihood to many thousands of people in Virginia and surrounding states.

And lastly, we tell of the ongoing comradeship of this team, even today, as it moves through the sometimes painful changes from the provincial paper mill of yesteryear into a global concern of tomorrow.

Foreword

This is a story of the Woodlands Division of Chesapeake Corporation, which was founded in West Point, Virginia, in 1918 by Elis Olsson. A Swedish immigrant who understood the art of pulp making for paper, he was successful in turning a small existing pulp mill into a company that has now grown into a global Fortune 500 tissue and packaging company.

For readers who may not be familiar with the Virginia Tidewater area, West Point is a small town about 30 miles east of Richmond, nestled amongst the Pamunkey, Mattaponi and York Rivers, off the Chesapeake Bay. It was and is the perfect setting for a paper mill.

This book project, started in 1996 in order to record some of the special oral history of the original Woodlands Division, is based on the interviews of approximately 75 past and present employees, wood dealers, loggers and others who have done business with the paper mill and make up a part of its long and interesting history.

Nothing is static in today's corporate world. As the book was being written negotiations were going on for the sale of the paper mill. In May of 1997, Chesapeake Corporation sold off its paper mill in West Point to St. Laurent Paperboard, Inc., a Canada-based company, and the book project was temporarily halted. At that time, the Woodlands Division was divided into two parts, the Marine Department and the Keysville sawmill going to St. Laurent; and Woodlands, three sawmills and the forestry staff staying with Chesapeake.

By 1998, the interviews for this book began once again. But something very interesting from the standpoint of a writer had taken place. There was a distinct difference in spirit, attitude and mood from the "pre-paper mill sale" to the "post-paper mill sale." The sale made a big difference. It was as if the heart had been carved out of the Woodlands Division.

But that was not the only trauma this book faced. In the winter of 1999, Chesapeake announced the sale of the last remains of the original Company once founded by the Olsson family. All the Company woodlands would go to Hancock Timber Resource Group. The last remaining three sawmills would go to St. Laurent. And the existing staff, many who had spent the greatest portion of their lives working for Chesapeake Corporation, would finally close up shop, turn off the lights, and go off to new jobs or go home.

And so there is a measure of sadness as this book is finally completed and the last words are typed and prepared for press. It would be impossible not to face these feelings or to present a book without this emotion. And it would be unfair, too, for it would not be honest. No book is worth writing or worth reading if it does not confront truth.

We are sad as the beginnings of this great Company are split up and divided like loot among pirates. We are unhappy too that the Company that we so loved and worked so hard for has passed on to nothing more than memories. But for us to leave this book on such a note would hardly be true to who we are or to what we believe. For we are Chesapeake people, born out of a Swede's dream in a small town in Virginia on three rivers and with the old man's heart and soul bred in our bones.

We are a positive, proud, strong, and a very special people who are never defeated no matter what might face us as we journey through life. And we know now, just as we know every word in this book, that this sad time in our lives is just the beginning of a new and exciting chapter.

So this book now ends and goes to press as this 20th century comes to a close. But the spirit of the original Company we knew and loved lives on forever.

Chapter 1

A Trip on the Tugboat "Sture"

How could I be so lucky? One would have thought I had just won the jackpot the way I danced around the house. The telephone call had just come in from Tommy Callis, marine superintendent for St. Laurent Forest Products in West Point, Virginia. The year was late 1998. It was all arranged. I was going on an overnight trip on the tugboat "Sture"!

For years every time I had crossed the Pamunkey River bridge in West Point, I had seen the "Sture" along with her twin the "Elis," Chesapeake's newest tugboats (named for the old man, Elis Olsson, who had founded the Chesapeake Corporation of Virginia and his son, Sture Olsson, who had also been CEO for many years) moored at the dock at the Chesapeake paper mill. That is, I saw them if they were in port. But many times, as my eyes swept the waterfront looking for the little tugs, the docks were empty, for most of the time the "Sture" and the "Elis" had been out hard at work moving barges and bringing in oil, pulpwood or chips to the paper mill.

But my eyes always searched for the "Sture" or the "Elis" as did the eyes of many others who lived in the West Point area and crossed the bridge through town during the normal comings and goings of the day. We looked for the tugs, couldn't help it. We thought of the tugs as family. Our gaze turning to sweep the shoreline along the Pamunkey River was as natural as the sun rising and setting over the river each day.

We wanted to see the tugboats sitting pretty as ever dressed up in their red, green and black paint. They symbolized Chesapeake Corporation and the Olsson family. Chesapeake was our Company, we felt it deep in our bones, our Company founded right here in our state of Virginia in the very town of West Point. And we were proud, mighty proud, that a company had been sired right here on the banks of the Mattaponi and Pamunkey rivers. We had helped build this Company with our own sweat, blood and tears and this Company had given us our livelihoods for many generations, since its beginning days in the early 1900s. This Company meant everything to us.

But times had brought big change to Chesapeake as to every other paper company across this nation and this change had hit the town of West Point in similar ways. The paper mill had been sold off by a board of directors, sitting in judgment in Richmond in 1997, to St. Laurent Paperboard Corporation, a Canada-based company, just as if she were an old horse that a farmer thought was of no further use.

We knew this was but the beginning of a new era that would spell tremendous and continuous changes to Chesapeake and West Point and her families and that there was nothing any of us could do to stop these changes or to protect ourselves from any harm these changes might bring.

But we could write a book. In the spring of 1996, Jim Vadas, then operations manager and Woodlands' right hand man to Jack King, called me to begin plans to do that very thing. Yes, we would write a book about the earlier days when the paper mill was Chesapeake and the people of Woodlands stood in all their glory as an integral part of the mill. This book would tell our story about how things were and this story would never die, no matter how many times the old mill at West Point might change hands.

I thought of all these things as I packed my overnight bag for my trip. I had already decided in my mind's eye that I would start the story about the history of the Woodlands and its people with a ride on the "Sture," a trip taken by so many Chesapeake pilots and crews up and down the rivers of Virginia and across the bay to the Eastern Shore, the land and waterways that interconnected that area known as "Chesapeake country."

I loved the tugs because they had all the qualities I admired in human nature. They were spirited, unpretentious, hardworking, dedicated, and they held on to their rugged position at sea like a bull dog guarding a bone from the rest of the pack. They never gave up until the job was done. No matter how rough were the seas or how difficult was the task, a tugboat never gives up.

These were the very same qualities I had found in my many interviews of past and present employees of Woodlands from the ax man in the woods all the way up to the CEO of the Company. What better symbol for the book and the Company than the tugboat "Sture"? If I could have my way, the "Sture" would even end up on the cover of this book. (You can see with a quick check to the cover just how much power and influence your author had!)

During the spring of 1997, when the mill at West Point had been sold, I had witnessed the big sale and also the terrible outrage and pain as the old Chesapeake was torn asunder. In fact, the sale had shut down the book project for over a year waiting for emotions to settle. It was a long year and a half, waiting for the go ahead sign with this book ever bubbling in the back of my brain.

But I already had most of the story, the heart and soul of the men and women of this great American Company down on paper. And I wanted to tell the story. The perfect touch, the Chesapeake symbol of the tough tugboat that never quits even in the roughest seas, the very start of my book would be a trip on the "Sture." All I needed was a call from Tommy Callis OK'ing my request.

It had only taken me three years to get permission to take a trip on the "Sture"! Was there some hidden resistance to having a woman writer nosing about on a tug? But authors are tough customers, even for paper mills. I, too, could dig in like a bull dog. No tug trip, no book.

The day finally arrived in November of 1998, just two weeks before Thanksgiving. My father and husband, who accompanied me on the trip, and I drove to West Point on a recent Sunday morning and waited for Callis in the visitors' parking lot. He picked us up and took us down to the dock, but not before noticing all of our luggage lined up on the street and ready for the trip. He took one look at the luggage and laughed. "How many suitcases do you need for an over night on the 'Sture'?" he asked.

We hurried down to the dock. The "Sture" was just coming in to pick us up, chugging through the Pamunkey River bridge as nice as pie with a big load of wood chips. I stood and watched her like a mother watches a child. Rugged, 87 feet of no-nonsense steel, she sported a black hull, bright green decks, three layers of red cabins, white trim, and was topped off by a pilot house that stood 30 feet off the water and a 10-foot mast. I swear she was the cutest little tug I had ever set eyes on in all my life!

First step was to meet the skipper, Captain Dan Bohannon from Mascot, Virginia (The regular skipper is Deltaville's Captain Bo Traywick who was off duty this trip). First mate was Bob Mercer from Hayes, Virginia, and the two engineers were Steve Healy and Doug Moskalski, who also served as cook. His sister runs the Blue Fin Restaurant in Gloucester,

Virginia, so good cooking must run in the family.

We were scheduled to make a fuel run with an oil barge to Newport News, Virginia, to pick up 9 thousand barrels of number 6 oil, black and thick as molasses. That load of oil would be used as a fuel source at the mill for an entire week.

The first order of business was to move a few barges filled with wood chips. The "Sture" fastened on to the barges amid ship and whipped them out from one berth, kicking up mud along the shore, and over to another mooring as if she were moving tinker toys. "Those chip barges hold the load of 50 trailer trucks," Bohannon told me as

Author Mary Wakefield Buxton and first mate Bob Mercer head down the York River on the tugboat "Sture." This trip sparked the book.

I looked over the piles of chips on the barges peaked together like a mountain range.

Before long the cook called us to the galley for lunch. Hot club sandwiches, Greek salad, chips, sweet pickles, and homemade bread pudding which was packed full of fruit. "If you think this is good," Healy said as he polished off his pudding, "wait 'til you see dinner."

We were soon passing through the Pamunkey River bridge with the traffic stopped on both sides and sea gulls exploding at our stern. The empty oil barge on our nose was fastened down hard with a rope as wide as my fist and two steel cables to our stern. We had perfect control of the barge.

An inspection of the engine room showed us twin GM 2300-horsepower diesel engines that had the power to push and tow two barges at a time if necessary. The deafening roar of the engines kept all conversation to a shout and finally led to our hasty retreat.

We were soon hauling down the York River at eight knots with the November sun dancing on the river. At such speed we kicked up a 50-foot swath of lacy white foam that lay like ice on the blue river as far as the eye could see. The cook had given me some pancakes left over from breakfast and I enjoyed throwing them to the hungry gulls who caught them with a squawk right in midair. Then I sat down on the stern between the steel cables with my feet up on the transom and watched the mill disappear to a wisp of white steam on the horizon.

"You're the first woman I know who's been on this ship," the skipper told me. That theme chases me through life like a shadow.

I laughed. "I bet I'm the first of many more, too," I said, and then we both laughed. Times were changing and we both felt it. "How do you like working for the French Canadians?" I asked. "Are you speaking any French yet?"

He laughed. "I called up the main office in Canada last week to ask them about my stock purchase plan and they were all talking in French so I had to hang up the phone!"

I looked over the equipment on the "Sture." The cable line and winch occupied most of the stern area with the big steel lines wound onto a reel just like a fishing line. There was a Japanese made radar revolving on the mast, a signal light, auxiliary control station, and what appeared to be a life raft in a box. The old equipment was blended with the new. An ax, which must be useless on a steel ship (except possibly for cutting rope lines), a brass bell once used for fog hung off the pilot house, and a life ring all reminded me of yesteryear. Meanwhile the telephone, radios, television screen showing four points in the engine room, automatic pilot, running hot and cold water, fully equipped galley, sonar, radar, showers, water fountain, heat, AC and electric systems all capable of running on generators reminded me this tug was state-of-the-art.

We had the York River to ourselves almost all the way to the Coleman Bridge between Yorktown and Gloucester Point, Virginia, where we finally saw some other boats. It was an unusually warm day, a temperature in the mid-60s, and the sun shining down on our faces was strong enough to give us a November sunburn. At Cheatham Annex, a Navy base dock where it is not unusual to spot an occasional supply ship, we saw what probably was a white hospital ship and a Naval transport ship but the Naval Weapons Station dock at Yorktown was empty.

The first mate, Mercer, had taken command on a six on, six off schedule. He lined up on a span with 60-foot clearance to pass under the bridge and almost immediately a pleasure boat began to turn across our bow. Mercer blew his horn in warning. "Once I'm lined up in the channel on a span with my barge, the pleasure boats need to stay clear," he said. After several toots from us, the pleasure boat got the message and moved fast.

We passed under the bridge and I looked up and saw the steel grating and the cars passing overhead. It seemed to me the roar of our engines was even greater under the bridge, as if captured and magnified by an echo under the steel. I looked out to the mouth of the river with the Yorktown Oil Refinery spitting up fire on the right hand coast. The Chesapeake Bay awaited.

I glanced over to shore on the left to notice my sister Alice's house in Bena, Virginia, a small community along the river. I wondered if she were watching the "Sture" in her glasses as she had a daily habit of watching tug traffic chug by from her kitchen.

We soon rounded "two marsh light" at the mouth of the York River and headed south, then slipped through the small ship channel off Grand View past the salt ponds and Buckroe Beach near Hampton, Virginia, where my husband, Chip, used to sail when a boy. The sun was hanging low in the sky, a bright red ball of fire, and the sea was as smooth as a piece of grey silk. I could detect a slight roll of the ocean even from my pilot house post 30 feet high.

The "Sture" had been built in 1979 in Louisiana. The 70-mile winds and 12-foot seas had been so rough bringing her up the coast from the Gulf that the pilot could put his hand out the window 30 feet up and feel the crashing waves break in his palm. One of these giant waves threw the pilot out of his seat and when he landed he was knocked out cold. Steve

Moskalski, the same crew member with us on this trip almost 20 years later, had to carry the unconscious pilot off to his bunk. That same storm had knocked the brand new 2000-pound stainless steel stove in the galley clear across the room and back.

No cruise ship was the "Sture." She was a rugged work ship, not equipped for fun or leisurely life. I could have tripped over a dozen objects used for some task on every deck.

I looked around at our facilities. The "stateroom" was a couple of bunk berths in a space no bigger than a walk-in closet with a porthole that could possibly lend a queasy soul a breath of fresh air. The stairways leading up to all three decks were steep and dark at night and I was grateful the waters were calm for our trip. Getting on and off the tug and later up onto the oil barge to investigate the workings of refueling required both muscle and energy.

As we neared Hampton Roads, we noticed many fishing boats huddled together at various favorite fishing spots. One tugboat captain had turned off its engine and was drifting along with his fishing line overboard. Every time he caught a rock fish he radioed over to us. "Hey there, tugboat 'Sture,' I just caught another one!"

There was some cross banter between the two tugs as to why we did not stop and go fishing' too. But we had work to do and a schedule to meet so there was no fishing' for us. We spotted the Chamberlin Hotel in Hampton to our right and several aircraft carriers docked at the Naval base in Norfolk on our left.

The sun, now hanging low in the west, appeared to be looking for a place to set. She chose the Newport News Shipyard for her nest, as nice a place to set as anywhere else. She slipped into the horizon, a huge dark crane silhouetted against her last blaze of light. She bid us fond farewell, a last molten ball of liquid fire, sending sparks out to sea in a glorious, golden path.

We steamed over the Hampton Roads Bridge Tunnel and headed for the oil dock in Newport News just as dusk arrived. Our skipper was soon turning the tug and barge on a dime in the cramped space of the oil dock harbor.

"The S. S. United States used to dock over there," Bohannon said, pointing to some big docks to our right. "There was nothing like the United States to give a tugboat some good leeward protection in a big wind," he added with a smile.

It was dinnertime and the cook called us to the galley in shifts for a meal of hamburgers, gravy and onions, string beans, corn, potatoes and homemade biscuits. I was amazed at how hungry I was. Everything looked so good.

"The food is the best part of tugboating," Mercer said. "You can tell right away if the cook doesn't like his job," he added. "There isn't anything worse than being at sea with a cook who doesn't like to cook!"

The cook was strict about one thing. We said a prayer before we ate. It lent a nice touch to eating at sea but I peeked at the steaming biscuits while I thanked God.

The crew untied the oil barge, tied her up at the dock, then pulled in beside her for the four-hour refueling. I climbed on the barge and inspected the fuel pumps to make sure everything was done correctly.

Soon, we tired of watching fuel pumps fill the barge. Even the gradual lowering of the oil barge into the water lost its appeal after the first hour. We retired to the galley which was the warmest spot on the ship. By now the temperature had dropped. The mid-60s November day had turned into a chilling upper-30s night. We pulled out a deck of cards and played a couple rounds of gin rummy with my lawyer husband beating me every time and then having the nerve to record the score.

The strong smell of fuel oil permeated the night air. "I'm going to bed," I announced, after losing five games in a row. I retired with a loser's flounce to my closet. I climbed into the upper bunk and listened to the many strange sounds, trying to imagine the reason for each groan, creak or clank. Later, I was vaguely aware we were positioning the barge back on our bow and moving out to sea. As we moved out into the bay, I fell into a deep sleep and dreamed of little tugboats passing in the night.

"Are you awake down there?" a voice startled me in the night. My eyes flashed wide open and stared into the black night. "Who is this, God?" I asked, ever hopeful.

"No, it's the skipper on the tugboat, 'Sture,' up in the pilot house," the voice answered. "Remember? You said you wanted to be on deck to see the mill when we come into West Point."

I did not jump out of bed and into my clothes like a crazed nut and fly up to the pilot house and carry on about what a beautiful sight was the mill at night. Rather I looked at my watch.

It was 4:30 a.m.! Please! No time for a lovely lady to be rising from her bed. "Lovely Lady-it is" hits me every now and then, but fortunately I manage to shake it right off like a bad cold. I thought about the spectacular sight that awaited me, coming into West Point at night with the paper mill lit up like heaven or the gates of hell, whatever one's inclination.

So I jumped out of my bed and into my clothes like a crazed nut and flew up to the pilot house carrying on about what a sight was the mill lit up against the black sky.

"It's Windsor Castle on fire against the black sky!" I shouted. "It's the king's birthday cake lit up with a thousand and one candles!" I continued. "It's the crown jewels in the London Tower!" I said, really on a roll.

The crew laughed. "We thought it was the paper mill," they said. Then, "Well, it is a pretty sight," they admitted, even though they had probably seen it a thousand times.

But I could have died on the spot with joy. I had to borrow a pen and flashlight from the skipper to record my images while they were still fresh in my mind.

Dawn broke as we approached the Pamunkey River bridge. Traffic waited as we slipped through the open span. Sodium gold lights of the mill and their answering stilt reflections shimmered in the grey glass river.

Father pointed to the west. "Look, the early morning mist is rising up like smoke off the marsh grass," he said. I looked at the sight that could have inspired a poet. Then I turned to look to the east. Some artist had painted a strip of pale pink on the horizon to usher in the sun.

The "Sture" dropped off the oil barge and sped back river to its berth next to her twin ship, the "Elis." We backed in to our berth with our engines alive like guns in November. I hopped off, baggage in tow, my sea legs wobbling as they hit land.

I waved to the crew. I saw their dear faces drawn with the hard work and fatigue of life on a tugboat. "I'll send you my book when it's finished!" I shouted over the engines.

Callis was waiting to greet us. He was all business. He checked to see if we were intact from our great adventure at sea, then sent us off for home. Nobody wants any writers hanging around after the show is over.

6

I blinked. It was Monday. My eyes had that scratchy, dry feeling one has after a bad night. As we left the mill, workers were filing into the mill with the stark looks of Monday morning etched in their faces. My husband had already changed into business attire and was headed to court in the nearby city of Newport News. His step had a roll to it and he walked off to his car like the ancient mariner finally come ashore.

Father and I loaded up my car and headed home for Urbanna, a town nestled nicely one river up in the Chesapeake Bay, a river known as the Rappahannock. I had a sense of deep excitement about me. I could still hear the engines on the "Sture." I could see the icy foam lacing the black river and I could hear the wild cry of the gull. I could smell the sea at low tide, oil from the tanker and even the hamburgers sizzling in the galley.

I could not have written a book about Woodlands without a trip on the tugboat "Sture." It just could not have been done. I learned from the very first day of my book project that there was one underlying and powerful river that flowed through this Company with the current of a thousand charged steeds. It was, simply stated, the absolute love and devotion of the Chesapeake people to the Olsson family.

It was also a fact that the tugboats in Woodlands were real symbols of bringing in the wood for all these years. My trip down to Newport News and back to refuel was a trip made hundreds of times by the past crews of the Marine Department. The tugs passed through rivers and creeks and bays that were all a part of Chesapeake country. My trip on the tugboat named "Sture" would symbolize everything I wanted to stress in the opening pages of the new book.

All the passion, hard work and devotion to duty that represented the old Chesapeake Corporation all were symbolized on the "Sture." The thundering engines, the boiling white foam, the fifty-foot-wide wake kicked up and spotted off stern all the way down the York River for miles and miles, the chugging along and getting the job done no matter what was the weather.

Oh, with a little bit of luck, I'll come up and down that river many times and I'll experience a lot more of life and I'll write a lot more books, too. But no matter what happens or where I go in life, one thing is sure. Let me make my position perfectly clear from the very first chapter of this book. Let the changes come. Let the future shine down on us as bright as it may or not. No matter what, my heart remains on the tugboat "Sture."

Chapter 2

The Old Man

It might be said that the tried and true, classic formula for the typical founder of American industry was a tough, swearing, hard working, myth making, highly spirited, hard drinking, risktaking white male immigrant who never quite learned to speak the English language without a heavy accent that revealed his recent arrival to this shore. If this is so, Elis Olsson, a native son of Sweden who immigrated to this country in the early 1900s and ended up founding the Chesapeake Corporation in West Point, Virginia, now a Fortune 500, global concern, fits into the formula about as perfectly as a stretch glove fits a hand.

And it did not take long for me to discover, as I worked on this book project, that the one underlying thread woven throughout every person and department in the original Chesapeake family, like the interlocking thread in that glove, is an undying love and admiration for Elis Olsson, his son, Sture, and the entire Olsson family. Throughout all the interviews that were conducted for this book, stories from the Chesapeake team about the father and the son and the Olsson family bubbled forth like boiling chocolate on the Christmas stove.

Everyone had a story to tell about the "old man." It was immediately clear to me that the most important fact a person being interviewed could transmit to me—and this was usually done right at the beginning of an interview—was, "I knew Elis Olsson." And if this fact could not be delivered, the next most important fact was, "I knew Sture Olsson." This writer never failed to notice and record the shining eyes and happy smile of every single employee who said so. It soon became clear to me that the men and women of Chesapeake were letting me in on some of the most important truths of their life. They had a special relationship with the chief. And they loved the chief and were loyal to him right down to the last moment in time and the chief returned both love and loyalty to them.

Thus, this book was built on this love and devotion. That needs to be said up front and right at the start. Also, this book is not a lot of historical facts guaranteed to put a reader to sleep or to lie on some distant library shelf collecting a lot of Virginia dust. This book is the real live story of real live memories of the real life staff who made up the Chesapeake Woodlands team.

I liked these people. For one thing, they came to me as honest as the sun on a new day in spring. They told me everything, up front, openly, and without a lot of hullabaloo. They did not try to hide anything from me, or pretend to be somebody they were not, and they did not have any hidden agenda. I like these traits in people and if my affection shows up in this book, then so be it.

Reader, think of this book as the woods. There are many trees in the woods and each tree is a little different from the next tree. But each tree is a very important component of the forest, for which tree can we really afford to lose? And each individual tree is dearly valued in the consideration of the whole.

How did the Company get its start? When did the Woodlands Division begin within the Company and how did it develop? For a detailed and factual account, readers are referred to a book written on the history of the Company, published in 1987, titled "Chesapeake, Pioneer Papermaker" by A.T. Dill.

But for a compilation of sweet memories, often full of fun, that came from the real peo-

Elis Olsson, Pioneer Papermaker

ple who worked in Woodlands and personally knew and worked with the Olsson family over many years, turn to these pages.

Special care was taken to record every single memory as closely as it was originally related to me as possible. In this way we hope to have captured some of the passion and joy and excitement that was and still is a part of Chesapeake Corporation and therefore present to readers everywhere a bit of very important and unusual history that can never be taken away from its people.

When it came to the beginning of the tale, I went right to the founder's son, Sture Olsson, for a recap of the Company beginnings. "Mr. Sture" sat in his office at Company Headquarters in 1996 for our interview. He was dressed in a suit and red tie, along with his usual bright red socks that can cause a writer to stare for one or two unguarded moments, with a favorite dog at his feet. The breed was a chocolate Boykin spaniel, a South Carolina breed, who, after checking me out to determine whether I liked dogs or not, spent the entire interview receiving long, luxurious strokes from me with one hand as I took copious notes with the other.

Fortunately for me, I love dogs. But my obvious adoration for the spaniel got me no noticeable favor from his owner. Mr. Sture looked me over suspiciously as I suspected his father might have done at one time if I had entered his office. Some damn female writer nosing about in my business and why in the hell should I talk to her, he might have been thinking. He might have done more than just think it. He might also have said it. (see appendix)

I went right to work sweet-talking the fellow to settle him down, so to speak, like women

9

do when they confront agitation, whether it comes in the form of a man, a horse, or a dog. I did not blame him one bit for his gruff manner. Who wants to speak to a writer who will only go off to her computer and record the conversation.

But as the interview got going, Mr. Sture seemed to relax and enjoy himself and then he began to tell me the good stories of the past. But he never once lost his suspicious demeanor. As he talked he was trying hard to size me up. Was I a friend or an enemy? I imagined it was this same character trait that had come down from father to son that had built this Company.

And then came his one final admonition to me. "Make sure I see every last word in this book before it goes to press!" This statement indicated perhaps another very important trait that caused Chesapeake Corporation to survive and grow to what the Company is today.

"The brainstorm behind building a paper mill in West Point, Virginia, actually came about under the leadership of Crosby Thompson from the Cleveland Plain Dealer," Mr. Sture said. "Thompson came to West Point at a time when the country was being pinpointed on maps by many concerns as to good locations for paper mills. After looking over West Point and the adjoining rivers and timberland, Crosby decided he had discovered a perfect site."

One of the first business moves Thompson made after arriving in West Point in 1912 was to purchase the local newspaper. "Soon after that," Mr. Sture continued, "he began to lay out the streets for a new city around the site for his mill. He then gathered some financiers from Cincinnati and an engineer from Boston, Charles T. Main, and built the first facility. Construction started in 1913 and it was completed the next year," Mr Sture said. But getting a pulp mill up and running in those early pioneer days was a complicated business. By 1918 it was still not producing even one pound of acceptable pulp.

The mill was consequently put on the market. "Soon Norway's answer to Aristotle Onassis, Christoffer Hannevig, a shipbuilding magnate from New York City, heard about the opportunity," Mr. Sture said. "He communicated to his friend and my father, Elis Olsson, who was at the time working as plant manager at a paper mill up in Canada. 'Elis,' Hannevig had said to my father, 'you ought to go down to Virginia and buy up that mill in West Point.'"

"Mr. Elis" did just that with Hannevig's backing and financial help. With his hard work and very special know-how in the paper-making industry, which was rare in America in the early years of this century, he finally had the West Point paper mill grinding out its very first sheets of heavy brown paper known as kraft paper.

As the years went by, Mr. Elis managed to take full ownership of Chesapeake Corporation and lead the young Company through the very leanest and most difficult times. "He liked to tell his friends and family how he started up the Company," his son remembered with a smile. "He told us, 'Mr. Hannevig talked me into giving up my job in Canada and taking a crap shoot in Virginia!'"

Over the years, many wonderful stories have surfaced about Mr. Elis by his employees who loved him just like a father. The stories about the old man tell a lot about who he was and why he was so successful.

One of the earliest memories came from Chief Webster Custalow of the Mattaponi Indian Reservation on the Mattaponi River not far from the town of West Point. Now in his mid-eighties, he still remembered riding down from the reservation with his father, Chief George

Forest Custalow, in the early 20s bringing a load of wood on horse and wagon down to the mill.

"In those days there was no fancy corporate headquarters," Custalow said. "Mr. Elis had his office located right in the mill so he could personally keep an eye on everything and make sure everything was running smoothly."

"Mr. Elis knew everyone in the Company on a first name basis," the chief remembered. "Later, this trait was picked up by his son, Mr.Sture, and folks used to say that once the father or the son met you, they never forgot who you were and always greeted you using your name whenever they saw you."

Consequently, when Chief George Forest Custalow and his son Webster arrived in front of the mill with a load of wood, Mr. Elis could look out the window and know exactly who they were. Only problem was one particular day when they had pulled up with a wagon load of pine, the Virginia State Police were also there. "They waved to us and told us to move our rig and not to loiter on the street," the chief said.

"Mr. Elis saw this and came running out of his office on fire as wild as a stallion," the chief said with a laugh. "He shouted to the police, 'Don't you dare try to stop any of my wood dealers from bringing in their wood!'" It seemed then to the little boy who is now the old chief that Mr. Elis was so mad that "smoke was coming out of his ears.

"This is Chesapeake property and don't ever forget it!" Mr. Elis shouted, shaking his fist in the air. No one wanted to hang around for any more of that Swedish temper. The police disappeared as fast as butter disappears on hot toast. Later Mr. Elis told the old chief if the police ever tried to interfere with him again, "'just let me know.' But it never happened again that I can remember," said Chief Custalow.

The Chesapeake mill in its early days, taken between 1922 and 1924.

The original corporate headquarters, built in 1928-29. This photo was taken circa 1940.

Jimmy Sears, a longtime employee and retired safety director for Woodlands Division, remembered that the old man drove a big, black Cadillac with red upholstery and everyone in West Point knew who he was when they saw that Cadillac coming. "Mr. Elis stopped anywhere he wanted to stop in the town and never bothered about any silly, old no-parking signs. If he needed to run into the bank and the parking places were filled, why, Mr. Elis would just stop his car in the middle of the road and walk right into the bank. Nobody would think of trying to move it, either," said Sears with a laugh.

When Sears was a young man back in the 1930s, he remembered selling newspapers in town along with several other jobs that helped pay his expenses. "When Mr. Elis came by I would run a newspaper right over to his window and he would roll it down and take it," Sears recalled. "He didn't even have to get out of the car when I was on duty. And for my excellent, personalized service he always gave me a bright, shiny quarter! That was five cents for the paper and 20 cents tip for me!"

Although the paper mill made money as soon as Mr. Elis took over, there were plenty of lean times, especially in the early days of the depression when customers could not always pay for the kraft paper they had purchased, at least not as fast as he might have wished. Sometimes cash flow was so low, the old man could not meet payroll at the end of the week and this was serious not only to his workers, but to the entire economy of West Point. So he paid his employees in special wood tokens that were known as "Chesapeake chits."

The chits were used just like money in town and all around West Point. People bought food with them at the local grocery store, paid utility bills, rent and even made payments on loans at the banks. When Chesapeake was rolling in cash again, the chits were cashed in, just like chips to a gambling house teller after a poker game. The system worked well when cash was scarce. Many people in West Point still remember holding those early Chesapeake chits that felt, according to at least one observer, "just like a large wooden nickel!"

Once a group of workmen were standing on top of a pile of sawdust in front of the mill busy shoveling it into a truck. All at once, a bevy of quail landed on a nearby mound. Before the men knew it, a black Cadillac appeared out of nowhere and pulled up at the foot of the sawdust. Mr. Elis jumped out and ran to the trunk and pulled out his shotgun.

Bang! Bang! The gun reported and the bullets whizzed by the men so fast they did not even get a chance to duck or run for cover. One of the men thought he felt a bullet whiz right between his legs. Then the old man came running up the sawdust pile and picked up his booty. "Roasted quail for dinner tonight!" he shouted happily to the men with a wave. Then Mr. Elis threw the gun and quail in his trunk and took off for home in a cloud of dust.

Mr. Elis knew all his men and loved every one of them. He was known for keeping the Company going and the paychecks coming even in the hard times when earnings were lean. Mr. Elis did everything he could never to lay anyone off. Occasionally there was a sign at the front gate of the mill that read "Not receiving any wood today," but it did not happen very often. He spent his whole life energy seeing to it that the mill was open and receiving wood every single day.

But if Mr. Elis ever got a whiff of an idea that anyone was not absolutely loyal to him with 100 percent devotion, the ax could drop with surprising speed. One day Mr. Elis proved as much as he stopped to look over his baseball team known as the "Chesapeakes."

Chesapeake Corporation began to sponsor a franchise baseball team in the Piedmont League starting in 1932 as the depression was just beginning to settle in on the local economy. Baseball games added a lot of fun to the Tidewater area and the "Chesapeakes" played other teams in Durham and Raleigh, North Carolina, and Richmond, Lynchburg and Culpepper, Virginia, too. Over the years the Chesapeakes had some pretty good ball players on the team including pitcher Vinton Sutton, who had been to the Big Leagues and was the local team's special pride.

Mr. Elis enjoyed baseball about as much as anyone else and liked to go and watch the games. But once when he pulled up to watch the boys practice for some unknown reason one of the players who he knew worked for his Company made a disrespectful gesture to him.

"See to it that man is fired!" Mr. Elis told his driver and they drove off. Sure enough, the next day the man was fired.

Later, in 1940, when the manager of the team, Bill Gwathney, died in a drowning accident, the Company gave up its sponsorship of the team. The top professionals left for other teams and the Company offered jobs at the mill to the rest of the team members including Jimmy Sears, Dickie Haley, George Howard and Monroe Sours.

In the 1950s, when Mr. Sture was starting to take over the management of the Company and the old man was slowly fading out, Jimmy Sears remembers a special moment. It was the very first supervisors' meeting Sears ever attended with the Company and Sears was all eyes and ears at every word that was said.

"Mr. Sture got up and talked a long time about how wonderfully positioned Chesapeake was and how we had timber over on Eastern Shore and in Tidewater and Central Virginia and this was the closest timber to the giant market of New York City and because we could transport the wood at a cheaper cost than any other company, we could outsell any of our competition. Then in a fit of enthusiasm Mr. Sture added with great passion, 'Men, our most important asset is our timberland!'

"This was too much for the old man," Sears said. "He stopped his son's talk and slowly came to a full stand. 'Son, I must correct you about a very important thing you just said, even in front of all your employees.' The old man looked fondly at his son and then at the room full of supervisors. 'Never forget, son, the most important asset of this Company is our people,'" Sears remembered.

With that one simple statement the old man sat down. But Mr. Sture never forgot what his father had said and he used the same philosophy over the years he was in charge of the Company. With every business decision, he put his people first.

Sears further believed that every employee of the Company knew he would be called upon to make sacrifices and to work hard for the Company. But he or she also knew they were loved and cared for by the Olsson family, just like family.

Florence Johnson, wife of Bud Johnson, who was chief forester at one time for the Company, remembered that the original home for the president of the Company was just right of

Shortwood trucks line up at the mill (circa 1930's).

14

the mill. But in time, Mr. Elis bought Romancoke, a big plantation on the Pamunkey River which had once been the home of Captain Henry Lee's son, Captain Bob Lee. "Every Christmas the Olssons threw a big party and nearly everybody in West Point was there. Those parties were really something and no one who was privileged to attend ever forgot them," Mrs. Johnson said.

Dick Cartwright, Woodlands Division surveyor, remembered the old man well. "He was a big, tall, hardworking fellow who still spoke in broken English. We saw him every day because he came into his old office in the mill and made his personal rounds calling everyone by name."

Others remembered Mr. Elis as a big, energetic Swede who worked hard and cussed hard. He was honest as the day was long, a real salt of the earth type of person, and when he gave you his word, you knew you could count on it.

He had his two boys, Sture and Carl, working for the Company as soon as they were old enough in the summers when they were home from school. Even though they went off to prep schools and universities, they worked up from the bottom just like everyone else learning the ropes. Both sons identified with their father's hardworking values.

"Mr. Elis made sure he spoke to everyone he saw every single day and knew everything that was going on. He naturally knew good management techniques and how to make a profit without ever once attending a business school or reading a book on how to manage people or be an effective boss. He was a "natural." He took personal interest in his people and he cared about them and his people knew it and loved him for it.

Such bonding between the chief and staff almost seems like a fantasy in today's cold and distant corporate world. The rule was simple. In those days the Company was just like family and everybody knew everybody and people got along because they knew they had to get along to get the job done. Not a bad way to run a business.

Chapter 3

The Son

Sture Olsson was born in Richmond, Virginia, July 1, 1920, two years after his father founded the Chesapeake Corporation of Virginia. That same year, Chesapeake posted a net profit of $307,568, which was a huge sum of money in those days. The year was filled with a glorious euphoria and high hopes in the Olsson family for the future. What could make a young CEO happier than a solid net profit so soon after starting his Company and having a brand new son?

One of two sons of Elis and Signe Maria Granberg Olsson, Sture began his schooling in the West Point public schools and went on to graduate from Episcopal High School in Alexandria, Virginia, in 1938, a school befitting the son of a successful business magnate. It was at EHS that Sture would make many important connections that would help him grow his business in future years. As the nation was pulling out of the pits of depression in the late 1930s, Sture went on to the University of Virginia where he graduated in 1942 with a B.S. degree in mechanical engineering.

The old man believed in starting work at entry level in business and working up from that point, which was exactly what he had done. He made sure both his sons (Carl and Sture) had summer work experience at the mill starting at an early age so they not only knew the business from the ground up, but also knew everyone who worked at the mill on a first name basis. This was a formula that had worked well for the father and he wanted to make sure it would also work for his two sons.

After Sture's military service was completed, he began work at the mill as an engineer in 1946, whereas Carl had become a war pilot. The very next year Chesapeake was listed on the New York Stock Exchange for the first time, a huge leap in business success. Working his way up the ladder of command and working in literally every department of his father's Company, Sture was well prepared to take over the Company. In 1952, at just 32 years old, Sture was elected president of the Company and Carl became vice president of Woodlands. That very same year sales were $17.8 million and 100 shares of stock would have cost $3000 to a prospective investor.

In 1957, Sture made the most successful decision in his life. He married Shirley Carter, not just a lovely Virginia socialite but also a hardworking, highly dedicated, round-the-clock physician who spent many years of her married life caring for the people of West Point and surrounding areas.

There are many stories circulating in West Point of mill employees who were patients of Dr. Shirley Olsson. On many occasions they would make late night or holiday calls over to the family home, Romancoke, regarding sick children. Sometimes Sture would answer the phone. He would then hear who was ill before turning over the call to his wife who immediately offered to help. Sometimes this meant Shirley Olsson would hop in her car and drive over to the home of the employee, her black medical bag in hand.

More than once she was accompanied by her husband. It was unusual and still is unusual for a company town to not only provide jobs to most of the people, but also to provide medical care from the same household. Shirley Olsson's care and devotion to her patients over the long years has never been forgotten by the people of West Point.

Sture was a dynamic leader, as was his father. But as the years went by, Sture liked to complain that his family's Company was slowly but surely taken over by what he called the "bean counters."

"Suddenly numbers were the name of the game," Olsson said, "and every company in the country had to produce the right numbers for the quarterly reports to the stockholders. If the numbers didn't work out right, there was big trouble."

Because numbers had to be right, Olsson said it was a natural turn in management to allow their accountants to run things. The problem was that many of the accountants, who suddenly found themselves in charge of much of American business, knew almost nothing about the products they were charged with manufacturing. Consequently, Olsson believed, the quality of American products was destined to go down, even if the numbers or profits stayed at the correct levels.

Sture Olsson and Bart, his trusty sidekick, upon Olsson's retirement as chairman of the board in 1994.

In 1967, the Board of Directors asked Olsson to step down as CEO for medical reasons and to serve the Company as chairman of the board. Olsson reluctantly agreed to do so and was succeeded by Lawrence Camp.

"I had health problems at the time," Olsson said. "I had suffered a stroke, and although I did not really want to relinquish control of the Company, the board members did not support me. I knew if I resisted, it would have triggered a terrible fight within the Company. So I finally decided to put the interests of my family and the Company first and I made the move," Olsson said.

Olsson served many years on the Company board. In 1994, the son finally retired fully and Carter Fox became chairman of the board and CEO.

Perhaps one of the saddest changes Olsson had to witness was the move of Company Headquarters from West Point to Richmond in 1987. Tau Crute, retired director of land acquisition, said it well. "That move to Richmond stripped West Point of most of its community leaders. It was a blow to the town. It even affected church membership. West Point has still not recovered."

But Sture Olsson, who kept an office in the old West Point Corporate Headquarters until the mill was sold to St. Laurent Paperboard, Inc. in 1997, always dreamed of moving his father's Company back home to West Point. "It was pure fantasy," Olsson remembered

The Commonwealth
THE MAGAZINE OF VIRGINIA

Reprinted from the December, 1958 issue

Mountain of Logs
at
West Point

A view at Chesapeake
Corporation of Virginia
paper and paperboard
plant

Thomas L. Williams
Photograph

Mountains of logs at West Point. This view of the Chesapeake plant appeared in the December, 1958, issue of Commonwealth magazine. (Photograph by Thomas L. Williams)

18

sadly, "like saying if we could just get General Robert E. Lee back for a while we could finally beat the Yankees." Nonetheless, it was Olsson's fondest wish to see the headquarters return one day to West Point, the town that had done so much for him.

A third generation of Olssons worked in the Company. Sture and his wife had four children, a son, Elis, and three daughters, Lisa, Anne and Inga. After graduating from college, young Elis worked in the Company and was promoted to the level of starting up a new box manufacturing plant in Mississippi for Chesapeake. After St. Laurent came into the picture, however, the last Olsson left Chesapeake. The young Elis Olsson is now flying charter planes hauling relief and medical supplies to sparsely settled areas in Belize, Central America.

Many of Sture Olsson's friends and ex-employees remembered the going away party when the corporate offices were moved from the paper mill in West Point to Richmond in 1987. "Sture Olsson stood up," someone remembered, "with tears in his eyes, and turned to us and said something none of us who were there will ever forget . . . . 'Never forget, my friends, those who have the gold, rule.'"

Sture Olsson stated to me quite openly that "the bean counters beat us. Tom Harris, the Woodlands chief who came on board in 1968, came into the picture and built the best Woodlands Division in the country. But we had too much financial investment in the mill. We could not diversify fast enough," Olsson said sadly.

In the world of business, the numbers always count. They are either right or not right, there is no in-between. There is no turning back. Lee cannot return to beat the Yankees no matter how much anyone might want it.

But what wonderful memories we share of the son. Like the old man, he was full of steam, hard work and good fun.

When the day came that Sture Olsson took control of his father's Company, an interesting phenomenon came about. The very same devotion the employees felt for the old man carried over like magic to the son.

Sture was a lot like his father. He had inherited a lot of his father's passion for the industry. He had that same explosive personality and absolute undying love for his people. It was almost eerie. Every time the Chesapeake family saw the son, they saw the old man once again. Thus, when Sture Olsson took over the Company, it was an easy transition because no one really thought the old man had left.

One thing was certain about Sture Olsson. His people adored him and as the years passed, he instigated as many myths as did his father. Here are some of the special anecdotes gathered for this book and told by those who knew him and adored him back in the days of American corporate life when employees really cared about their boss.

Once, wood dealers from the Lynchburg area, Ralph Clements and Bob Sales, and sawmill and chip supplier George Ragsdale were attending a big party in Richmond and met up for the first time with someone called Sture Olsson from the Chesapeake Company in West Point. "None of us knew anything about him so as our conversation advanced, we finally asked him who he was and what did he do for a living?" Clements said.

"'I'm a wood producer,' Olsson answered as humble as pie. Later, when we found out this Sture Olsson was the owner and manager of the Chesapeake Corporation, which was only the biggest company in Virginia, we about fell over backwards!" Clements laughed.

Carter Fox, retired CEO of the Company, remembered that Sture Olsson once accompanied his sales team to a big convention in New York. Now Sture was well known back in

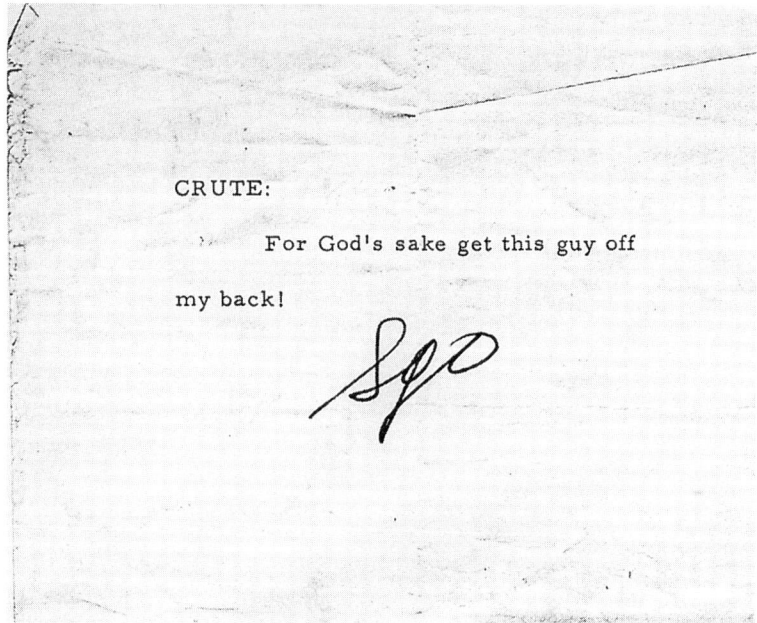

CRUTE:

For God's sake get this guy off

my back!

A note left on the desk of Tau Crute by CEO Sture Olsson (1950's).

Virginia for always wearing bright red socks. It made no difference to him whether he was dressed informally in a pair of old pants and a flannel shirt for a walk in the woods or in a three piece pin-striped suit suitable for a day of big business in the big city.

One evening Carter Fox was introduced to an important customer who looked him straight in the eye and said. "Say, your new salesman, that Sture somebody or other, is pretty sharp and I like him very much but you're going to have to do something about those red socks!"

Tau Crute, from land acquisition, worked for the Company back in the times when employees had a close relationship with the boss. He remembered flying with Olsson on a Company plane to view a horse race in Hot Springs, Arkansas. Sture was attending the race because one of his fillies was running.

"As we flew west, we passed a whole line of thunderstorms and tornadoes in our little plane," Crute remembered. "When we finally landed, it was pouring rain in sheets of water but miraculously, just before the horse race was to start, it cleared up and the sun even came out. As if the horse knew how much effort it had taken for us to get there and didn't want to disappoint us, she won the race," Crute said.

"The purse was $276,000. We were mighty happy coming home and we stopped in Lexington, Kentucky, to refuel and eat dinner," Crute continued. "I remember it was Easter and when we took off again for Virginia one of our engines malfunctioned and the pilot was lucky to get us down safely. Since it was a holiday weekend, all the commercial flights were booked. Sture just got on the phone and called over to West Point to have another plane pick us up. I never forgot what a nice feeling that was. It's great to fly with the president of the Company."

Jimmy Sears, retired Woodlands safety director, enjoyed a close relationship with Sture Olsson just like he had with Sture's father. Sears claimed the Olssons cared about West Point people and all their employees and employees' families. "They never got too fancy or successful to take an avid interest in anything that had to do with the town," Sears said.

"In 1941, I was representing West Point High School in the state final track events in Charlottesville," Sears added. "At that time, Sture Olsson was a senior at University of Virginia and probably pretty busy with his college activities. But as I was running the 220 low hurdle, I looked up and saw Sture standing in the crowds and cheering for me." Sears said he was so proud to have Sture Olsson there rooting for him. He never forgot it.

Sears remembered a conversation that demonstrated to him how close the Chesapeake Corporation family was during the Olsson years. "Every summer Sture would send his bird dog, Smoky, over to Ralph Clements' place outside of Lynchburg, Virginia, where the mosquitoes and ticks weren't so bad," Sears said.

One day Sears, who was attending a conference in Charlottesville, got a special call. "Jimmy," Sture ordered from his office back in West Point, "go by Clements' place on your way home and pick up my Smoky." Jimmy promised he would. "But don't you put my Smoky back in the trunk, ya hear?" added Olsson. Jimmy responded, "Sture, you can rest assured Smoky will be up in the front seat with me so close to me he could drive!"

Another classic story Sears remembered was one day when Olsson was a young man and still an engineer working during a period when M. C. McDonald was president of the Company, right after World War II. "Sture wanted to take a close look at the mill smokestack to see just how much tar and soot was building up on the tip so he called me up and asked me to fly him over the mill in a Company plane so he could get a really good look."

The smokestack was about 380 feet tall and Sears, who flew the Company planes and, at that time, was flying a two-seater, zoomed in about 500 feet over it knowing he had to be careful because the laws prohibited pilots from going any lower than 1000 feet. Sture had his binoculars and notepad but kept complaining to Sears he could not see anything unless they got a lot closer.

Now, Sears was the sort who definitely did not mind having a little fun and even with the old man's son. So he made another pass so close to the smokestack that Sture could have reached out his hand and touched it. "Damn!" shouted the future CEO over the roar of the engine. "Let's get the hell out of here before we are both arrested!" Whether Sture got close enough that trip to see any soot buildup, Sears never heard. Only he never had another request from Olsson to fly him over the Company stack!

Claiborne Courtney, who came to the Company in 1949 and was a long time employee in Woodlands, remembered the father, Elis Olsson, always had him come over to stay at Romancoke and chaperone Sture whenever he was away. Courtney remembered one night Sture had a girl friend with him for dinner. Courtney pulled out a detective badge he had in his wallet and pinned it on his chest. Then he marched over to the girl friend and said, "I'm an official detective, ma'am, and I'm going to check on you every hour, on the hour."

The wide-eyed girl promptly ran back to Sture and told him they had to deal with a professional detective who would be checking on them every hour! Sture laughed. "I know this detective. He's an employee and he better only check on us once!"

Robert Geron, a longtime employee from a family who had five members of its family at Chesapeake, remembered he and others used to grade Sture's driveway at Romancoke each year and other duties, including an annual burning of the marsh. They used to make a fun day of this task and turn it into a hunting party that Sture himself would join.

Another time Geron remembered Sture driving up to the boys in his big Cadillac, jumping out, opening up the trunk and unloading a big case of cold beer. "Damn it!" Sture shouted with an unhappy look. "My wife's got a meeting this morning and she told me I have to go and miss all the fun!"

One year, Sture's son, Elis, had a marsh fire out at Romancoke during a time his father and mother were out of town on Company business in Wisconsin. The fire was soon out of control and ready to do a whole lot of damage. Geron and his men showed up with a bulldozer and a fire truck to put out the flames.

After the fire was finally put out, Geron looked at the young boy sternly and asked, "Does your father know about this fire, son?" Young Elis somberly nodded his head yes. "And I didn't need a telephone to hear him explode either," young Elis reported. "I could hear him shouting all the way from Wisconsin!"

Paul Harper, retired Woodlands chief, remembered once Sture, Elmer Curfman and he were returning by car to the New Orleans airport from Huma, Louisiana, late at night. Curfman was driving and since it was so late both Paul and Sture had fallen asleep. When Sture woke up he found Curfman going around and around a traffic circle. Sture started raising hell that they were being driven in circles. "Well, I forgot what exit it is," Curfman said, trying his best to explain things. Sture promptly responded. "The first step in the right direction is to get off the damn traffic circle!" he shouted from the back seat.

Once Olsson and Harper were on a trip together in South Carolina to look at a chip mill operation. "The owner had agreed to meet us at the airport and take us to his country club for dinner. We had never met him before but as soon as we got in his car, it was immediately obvious to us he was a strong Clemson University supporter," Harper said.

"Right away Olsson started making comments about some 'cow college' down south trying to beat the great University of Virginia. Our host stopped the car on a dime and turned to Sture. 'Why, you S.O.B.! Say one more thing about the Clemson Tigers and you will eat hamburger tonight!'"

Sture enjoyed keeping race horses and one of his old time favorites was a five-year-old horse by the name of Rugged Bugger. He talked a lot about Rugged Bugger and one time he took his then nine-year-old son, Elis, fishing out west on the Salmon River.

"We were walking slowly upstream with our lines out against a big current," Sture recalled, "when all of a sudden a salmon struck on Elis' bait. The fish went crazy and my son was suddenly in a fight for his life. Following the axiom that states 'like father, like son,' Elis turned to me and said, 'Daddy, this fish sure is a rugged bugger!'"

The Olsson family to this day enjoys a loyal vote of approval from those who

Thelma Downey, former executive secretary to Sture Olsson, at a company picnic in 1993.

worked for the Company. They represented a day and time in American corporate life when CEOs loved their business and personally and passionately cared about their people.

Still every time I heard this sort of sentiment from a retired Chesapeake worker, I made a note of it. "In the old days when the Olssons ran this Company, none of us ever had to worry about being laid off," Carroll Dixon, a retired employee, said. "We had a union and if we had a grievance, we would just all sit down together and talk out our problems. If you ask me, the new people don't know that much. The workers today more or less have to train the supervisors. But the Olssons loved their people and always took good care of them. We won't forget it either," Dixon added.

Olsson now enjoys an active retirement at Romancoke and still daily comes into his office at the Chesapeake Forest Products Company Headquarters at the Claiborne Courtney Seed Orchard in New Kent County just outside West Point. There, positioned at his old desk, surrounded by the pictures of beloved members of his family, friends and many mementos of his long service to the Company, and usually with his dear spaniel Bart at his feet, he attends to correspondence.

And the foresters are mighty proud to have him, too.

Chapter 4

Early Days on the Waterfront

Since the very beginning of the Chesapeake paper mill in West Point, there has been a boat of some sort, manned by a rugged Chesapeake crew, that went up and down the rivers and creeks collecting pulpwood and bringing it back to the mill. In the early days Chesapeake boats were very different from the state-of-the-art, fully equipped, modern-day tugs and barges we see on the waterfront today.

When St. Laurent Paperboard Corporation of Canada purchased the Chesapeake paper mill in West Point in the spring of 1997, even though the Woodlands Division that once operated the marine department stayed with Chesapeake Corporation headquartered in Richmond, a very important part of the Woodlands went to St. Laurent in the deal. It was this very Marine Department which had been such an integral part of Woodlands from the start. For it was this department that carried out all the transporting of pulpwood, plywood logs, sawmill logs and chips from creeks, rivers and even from across the Chesapeake Bay at Pocomoke City, Maryland, and other points on the Eastern Shore and back to the mill.

Today, two modern tugboats, the "Elis" and the "Sture" and an entire fleet of chip and oil barges can be seen at the paper mill's waterfront managed by marine superintendent, Tommy Callis, a native of Mathews County, Virginia. Looking at this modern, hardworking fleet that contains all the latest equipment and comforts not even dreamed of in earlier times, like twin 2300-hp GM engines, radar, sonar, even a cellular telephone to check in with the wife and children back home, it is hard to imagine what the waterfront was like in earlier times. The first boats that served Chesapeake Corporation were much smaller and more primitive in every respect.

When retired dockmaster Otis Timberlake and retired tugboat captain William E. Oliver came to work for the Company in the mid 1940s, the fleet was very different. Oliver remembered that the first workboat the Company owned was the old "Ethel," a 50-foot boat with a draft of five feet. She was small enough to go up the many little creeks of Virginia rivers and pick up a load of wood and bring it back to the paper mill.

The "Ethel" would stop along the rivers and creeks just about anywhere there was a load of wood to pick up. She would sort of tiptoe in and out like a cat, pick up a load of wood, and chug back out again and all done with hardly any kick-up of mud. She was a perfect size for her task and thus the "Ethel" had many years of hard service to the mill in bringing in the wood.

Chief Webster Custalow of the Mattaponi tribe remembered the old "Ethel" and also the "Clarabelle," a sister boat, chugging up the river to pick up wood at his landing at the reservation. It was a big deal when the Chesapeake boats arrived. The pulpwood cutters and loggers knew they could sell their wood and pocket some cash every time the boats pulled in.

These early boats were followed by the "Atlas," a larger craft with a draft of seven feet and then the "Chesapeake," the first steam-powered tug which was later converted to diesel fuel at the Newport News Shipyard. Other early boats Oliver and Timberlake remembered were the "John F. Lewis," 105-foot-long barge with an 800-hp engine and the "Nichols," which was a medium size tow boat and the first of her kind at the mill.

Early "lighters" or motorized barges used by the Company were named for the rivers

Rows of wood at the Pittmans Cove Barge Landing near Kilmarnock, Virginia (circa 1960).

they worked in. Among them were the "Potomac," "Pamunkey," "Mattaponi," "Tollchester," "Ware," "York" and the "James River," which was a small World War II liberty ship.

The best known early boat was certainly the "Chesapeake," originally built in Camden, N.J. in 1889, and bought by the Company in the 1930s. Originally powered by steam, she was converted to diesel fuel in 1938 with the installation of a Fairbanks-Morse direct-reversible engine. The 77-ft.-long, 18-ft.-wide and 77-ton work boat added a lot of muscle to the growing Chesapeake fleet. This tug is now in private hands and is being fully restored.

The early boats never left the York River and surrounding creeks. They were very important to the Company because, in those days, they brought in almost half of the wood to the mill. They worked long and rugged schedules because the mill never closed and it needed wood every day, day in and day out, including holidays, regardless of the season or weather. Everyone in the Company and community understood one underlying fact of life. Without wood coming in each day, the mill would close. No wood meant no jobs. It was as simple as that. So the men and women devoted themselves to the gargantuan task of bringing in the wood each day to the mill, no matter what.

The boats would chug up and down the river in all weather, nosing into this creek or that cove, covering all the surrounding woodyards, picking up loads of wood ready for transport to the mill. In those days pulpwood was cut by hand by the old crosscut saws or by ax and later gasoline-powered saws. The wood was then lugged over to the boat by hand or homemade wheelbarrows or sometimes the logs were rolled off the bank down a chute and onto a barge.

Loading an old "lighter" with wood to go to the mill (circa 1920's or 1930's).

The Mattaponi River had nice, high banks perfect for rolling logs over and onto the barges, which really helped the loggers who had to use their own manpower to deliver wood. Custalow once asked Elis Olsson why the mill didn't have any landings on the Pamunkey River to collect wood. "No high banks," the Chief remembered Mr. Olsson told him. The early boats had none of the fancy equipment used today to help navigate the creeks and rivers. Timberlake remembered in those days the pilot had to run without radar and in fog, mist or storm conditions. This could mean running along at a slow speed using the whistle or bell and even sometimes stopping the engine and drifting dead in the water to listen for the echo just to figure out the next bend in the stream. "Some days the fog on the river was so thick you couldn't even see your outstretched hand," Timberlake said.

It must have made for some pretty scary trips up those rivers and creeks because the mill could not wait for clear weather to collect wood.

One can see an old picture of the tugboat "Chesapeake" and imagine her drifting dead in the water on a foggy morning as the captain tried to ascertain where to make his turn on the river. Perhaps the skipper tooted the whistle and then all crew members strained their ears to pick up on the echo. One can almost see the cook standing on the bow with a galley apron still tied around his waist, then hear him shout out to the skipper in the cold morning air, "Turn to, Captain! The bend's dead ahead on portside!"

Worse, it could be in the middle of winter and the days would be frigid cold and sometimes the decks would ice up and the men would have to work on floors as slick as oil. The ice could get so thick on the deck that the only thing the crew could do was take an ax and chop away as much ice as they could so they could get a foothold every now and then. It was cold out there on the river in the winter and the men remembered how the "night wind could blow icy knives clear to the bone." And even a simple act of getting to and from the bathroom meant a freezing, sometimes even dangerous, trip along the outer deck.

The new tugs have cabins completely inter-connected by stairways and passages with heat and cooling systems that protect the men from extreme weather, a far cry from what life was like on the "Ethel" or "Chesapeake".

Captain Oliver experienced a lot of excitement on the water during his long years working on the boats. "One Christmas I was out in a tug towing a loaded barge and the winds built up to a 50-mile gale with seas so rough you had to tie yourself down in your seat," he remembered. "Then, wouldn't you know it, at Wolf Trap light my engine suddenly went dead! It was so cold that night on deck that my spit froze almost before it could meet the deck. Suddenly I looked up and saw that barge rolling at me so close I could have thrown a ball to her deck! I called the Coast Guard and they came right out and helped us cut the barge off before we collided. I finally got the engine started. It had water in the fuel. And we made it back safely, even saved the barge. In all our years on the sea, we never lost a tug," Oliver said proudly.

Oliver was once on the "Chesapeake" in winds that gusted to 70 knots. "The 12-foot seas were so rough my pilot house dipped into the sea every time a big wave hit us!" he said with an expression on his face that indicated he had not forgotten it either.

And no one could ever forget the trip up the coast with the brand new tugboat "Sture" back in 1979. "Tugs can take more than men can, you know," Oliver said with a grin at the memory of that journey. "But that trip was so bad the pilot we had hired to help us up the coast got thrown from his chair and knocked out cold. The shipyard had forgotten to bolt down the one ton steel stove in the galley. It was thrown clear across the room and back before we could lash it down!"

Although the Company never lost a tug, they have lost a barge or two. Timberlake was on a run coming down the Rappahannock River from Colonial Beach when a barge he was towing started taking on water. "She sank but we worked so hard to save the load of wood that when it was all over it turned out we had only lost 12 sticks of wood out of a load of over 1000!" he said.

When Timberlake started with the Company as a deck hand, he was making 72 cents an hour. By the time he retired, he was up to $18 an hour. "It was hard work, too. All I remember all those years is pulling on rope, pulling on rope, and pulling so hard that your tongue hung out."

The "Chesapeake," the oldest working tugboat on the Bay, is pictured during the 1970's. It is no longer in service and is being restored by a new owner.

The tugs and barges today are constructed of steel and fairly easy to maintain but the early fleet was built entirely of wood. Maintaining wood boats was always a lot of trouble, especially in salt water environments where every single piece of wood must be properly refurbished and in a timely manner. Timberlake was in charge of many of the repair projects on the dock.

"It was bad sometimes working up in that crane with the wind blowing 45 knots and ice built up along the docks," he remembered. "Once I called the boss at his office and told him I would have to come down and wait a bit for the wind to die down before I continued with my work on the crane," Timberlake laughed. "And I never will forget what he said. 'I don't know what you're talkin' about, Otis,' he told me, 'it ain't blowin' down here!'"

The tugs and their souped-up engines go full speed between barges as they nose them in and out of berths or push them up and down the rivers almost every minute of the work-day. The tugs are tough, they work hard, they never quit until the job is done and they almost never are left resting at the dock.

Forester Dick Brake remembered a tale that was told by a couple of the boys on bull-dozers about a time they had decided to challenge the Marine Department. "Our bulldozers are tough, too," one of the men bragged. "We can down a 100-foot tree like a snap of a twig."

The Marine Department scoffed at the idea of a "tough" bulldozer. They decided to take the bulldozers on. The deal was set. A chain was connected between the tug and the biggest bulldozer the Company owned. Somebody blew the whistle. The pull was on.

The tug's engines roared and the husky ship kicked up some bottom mud in the river and a swirl of foam. The bulldozer answered with the same deafening response, her huge steel body screwed to the earth and roaring like a bull. For a while there appeared no movement from either party. Neither beast would budge in the great pull.

Then the tug, quick as a wink, turned sharp to starboard in the sea. The bulldozer flipped over on her side like a pancake. There was a whoop of laughter from the men. The challenge was over. The battle was waged and won. The tugboat was still king of the Company.

Oliver remembered piloting tugs in the early days without government licensing, which did not come into effect until 1973 when all the men had to get their "official" licenses. It was tough work and the crews worked around the clock. "But the company took care of us," Oliver said. "My wife had 21 major operations and Sture Olsson took care of 95 percent of every single bill."

Over the years, Oliver thought the biggest challenge he had to face in his work was adjusting to the changeover from the traditional eight-man crew down to the four-man crew. As the tugs became more comfortable with modern equipment to help do the work, there was less need for large crews.

There was an occasional debate between the union and the management over salaries, benefits and work conditions. Oliver remembered once the union called a strike and there was an all night session of hard bargaining. "The union said they wanted this and that and then Dan Lewis, vice president of Administration, came out and told us, 'Boys, that all sounds good but you don't want the deal because we can't afford to give it to you and therefore you can't afford to get it!'"

Timberlake remembered once when he and his union negotiated all night long until 4 or 5 a.m. in the morning with Tom Harris. "We were trying to win a raise of our expenses of five cents per meal per man," Timberlake laughed. "It was tough going and we argued like it was big money and we wouldn't give in either! Finally we all got tired and said we wanted to go home and go to bed. Harris agreed and gave us the deal," Timberlake added with a chuckle.

Over the years, the union and management at Chesapeake have gotten along amazingly well, which sets rare precedent in American industry. Perhaps this is because the Company had so many multi-generational families working together and everyone knew everyone else so well. And there was family input between relatives, too. Who wanted to go on strike if a father, brother or a cousin would end up out of work? Of course, the present chief of Woodlands, Jack King, ended up marrying Otis Timberlake's daughter. This is perhaps just one more reason everyone worked so hard to settle Company disputes peacefully.

Chapter 5

The Mattaponi Indian Connection

It could be said that the Colonists would never have won the Revolutionary War against England without the help of the Mattaponi Indians. It might also be said Elis Olsson could never have made a success of his newly formed Chesapeake Corporation without father and son Chief George Forest "Huskanawanaha" Custalow and Chief Webster "Little Eagle" Custalow and the Mattaponi Indians who spent much of their lives providing wood to the mill.

The original paper mill in West Point in 1914 was known as the West Point Pulp and Paper Company. Chief Webster Custalow remembered, "It looked like an old coffee pot sitting in the marsh with its one great big stack standing up tall." He saw this vision many times as a little boy accompanying his father back and forth between the mill and the Mattaponi Indian Reservation in a horse-drawn wagon.

The old chief and his son Webster delivered wood to the original paper company until it folded in 1916. The mill never did turn out any paper, and they waited six months to receive payment when the company went down. Even 80 years later the son remembered the long wait to get that payment and how important it was to the family when it finally came. Those memories show even today how necessary wood was for the livelihood of many families in the Tidewater Virginia area.

Later, Chief Webster Custalow remembered the high hopes his family on the Mattaponi Reservation had when Elis Olsson first came to West Point and took over the little pulp mill and formed the Chesapeake Corporation. One of the best sources for quick cash for the chief and his people was the sale of wood which was available in such plentiful supplies on reservation lands along the Mattaponi River.

At the beginning, no one knew for sure if Elis Olsson's new papermaking equipment would work and it took time before the new paper machine could get up and running. Who knows how much pine pulp went into the process of trying to produce the first kraft paper and paper board, that heavy brown paper used for bags and boxes still made today at the mill in West Point.

The first sticks of wood that went into Elis Olsson's new pulp mill came from Indian woodlands directly behind the present Mattaponi Museum. Chief Custalow said he thought his father at one time was sole wood agent for Elis Olsson.

The old chief and his son Webster "Little Eagle," who became the present chief, were as close as blood brothers to Elis Olsson. "He knew he could count on the Custalows to get the wood he needed into the mill every day of the year regardless of weather conditions to keep the mill open," the chief said. Without that kind of devotion to duty, hard work, and sacrifice, Chesapeake Corporation would have never become what it has grown to today.

Chief Webster Custalow was born in 1912 and remembered well those early days and his family's close connection to the Olsson family. He recalled that as a teenager he used to help his father haul wood on horse and wagon. "A round trip back and forth to the mill from the reservation took all day just to deliver a half cord of wood."

Then in 1925, Chief Custalow's father purchased a model T Ford truck and he let his son drive the first load of a half cord of pine into the mill himself. In those days Webster and his father and brothers chopped down every tree they took to the mill with an ax. They cut the wood into eight-foot lengths, later changing to five-foot lengths better suited to trucks. They even carried the wood by hand to the wagon or truck, then drove it themselves to the mill. Later they used a one-man "buck" saw and then in the mid 1940s they used a two-man power saw weighing in the range of 90 to135 pounds. Custalow and his crew were strong men and worked hard every day of the year in order to meet the needs of the mill.

Chief Webster Custalow, who is now in his late 80s, grew up on the reservation in a tribe that he believes is one of the earliest tribes in the country, dating back to 1652. He graduated from King William High School. After high school, he was lucky to get a job at the local pickle plant in West Point where he worked for five cents an hour or 50 cents a day. One day the bossman told him he was such a good worker he would put him on salary for $6 a week. He was thrilled with his raise. "Trouble was, I soon discovered the new 'week' turned out to be two weeks long!" the chief said with a laugh.

Chief Webster Custalow is shown in his home during 1996 with a favorite pet.

Later, Custalow bought a few trucks himself and started working for the Virginia Highway Department building new roads in the area. With his trucks, he was able to become a wood agent and Cecil Woodward from Chesapeake hired him as a wood dealer in 1932. "Cecil always told me I hauled in more wood for Chesapeake than any other agent," the chief added.

When Mr. Red Highland took over at Woodlands, the chief became his right hand man and continued his service bringing in the wood. The chief's hardworking family tradition kept wood coming into the mill even when "the snow was deep on the ground and the river iced up so thick the Chesapeake boats couldn't get through to pick up the wood."

"You know, Indians are tough," Custalow said with a wise old smile. "We can work in the woods in all sorts of weather and don't call in sick, even when it turns bitter cold."

Bringing in the wood in the 1920's.

Chief Custalow said it was his personal duty to see that the mill never closed for lack of wood and over the years many crews worked for him and helped him meet his commitments. Sometimes his crews worked on and on, even when no one else could get a crew out on the road.

One time, he remembered, Mr. Woodward called him up and the weather was so bad he said, "Chief, if you can get men out working in the woods today, tell them I will personally give them each a pair of new boots, compliments of Chesapeake Corporation." Woodward then called up George Ashley in West Point who sold boots at his store and bought every last pair of boots in West Point that day.

"But my men came out and they always came out and they worked hard," Custalow said. "The wood measurer came out every morning to measure wood at 7 sharp in front of the mill and my men were there at the head of the line every morning at 6 sharp. That was the way we did things, always worked hard, always did what we said we would do, all to keep the Company going because without the mill, West Point would have had no jobs and no money."

The chief remembered when a cord of wood only brought in 35 cents and men would chop wood for five cents an hour. "When I first went to Chesapeake, Woodward paid me $4.25 a cord and I would pay 75 cents to the landowner for a cord of wood. When I had to hire people, Chesapeake helped me meet my expenses. Those were the days. Of course, you could fill up a model T truck for $1 for gas and get a dime change, too, and just $3 filled up the truck with a whole load of groceries."

The chief and his men developed a loading system on the river to get the wood down to the mill through the use of "chutes." They were built on the high banks and when the men hauled the wood to the landing, all they had to do was push it down the chutes and onto the

This is the way it was done in the old days (circa 1940's). Wood is stacked and drying prior to being loaded on trucks.

barge where it would be transported to the mill.

The chief remembered many hard days in the woods. "The winter of 1936 was so bad the river froze up and the ice was 25 inches thick. The boats couldn't move so we got out our trucks and we worked day and night and the motors never cooled, we worked so hard," the chief said.

Then men worked with simple tools like axes, one-man or two-man saws and wheelbarrows. And those wheelbarrows weren't the bright red, shiny steel sort we see today but many were crude, handmade, really rickety wooden wheelbarrows that occasionally can be seen in someone's shed even now, 50 or 60 years later.

During World War II, the chief and his men worked German prisoners of war at Sandy Point Landing. "The Germans were good workers and they worked hard and all they wanted was a pack of American cigarettes at the end of a hard day's work in the woods," Chief Custalow said.

In those years Mann Bland, Chesapeake's wood measurer, would come over to the landing every week and personally measure all the cut wood in the racks. "I remember how he would mark the ends of the measured wood with a piece of red chalk to make sure he wouldn't measure the same wood twice and then he would figure up how much he owed us and paid us right on the spot," Custalow said.

33

The big tugboat "Clarabelle" could not get all the way up the river to Aylett landing so the Company brought in their little boats, the "Ethel" and the "Vamp," captained by George Gulasky and John Smith, two local men who were early Company pilots.

The chief remembered some of the first trucks his crew used in the woods. "The old 1933 Ford V-8 truck was fine on the road but it couldn't pull out of a ravine with a full load of wood! Eventually the trucks were built with more horsepower that could do the job," he said.

"By 1972, my arthritis and other health problems caused me to retire from the wood hauling business," the chief said. "Chesapeake called me in and took me over to a site they were working in Mathews County to show me all the new million dollar equipment that made the job so much easier. They offered me a million dollars worth of credit if I wanted to stay on. But I decided I was too old to take on that much debt and so I retired," he added with a laugh.

In all those years of bringing in the wood, the chief never saw more of an injury than a chopped off toe. "But I saw many close calls in the woods when the trees were coming down and they almost hit someone. I am very grateful for all the company did for me and all of us who live in this area," Custalow said.

Chapter 6

The First Foresters

As the 1930s arrived and the production of the heavy brown kraft paper at the paper mill was rolling along at growing levels of production, Chesapeake found it was finally ready to hire its first forester. Not only would a real live, authentic forester give the relatively new company a measure of well-deserved respect amongst its peers, other paper companies across the nation, but a forester would begin to set plans for a much needed forestry department with a long term program for growing and harvesting trees on Company lands.

Elis Olsson had started to amass Company timberland and many of his initial buys were tracts bought up dirt cheap during the depression. Most farmers were so hard pressed for cash during these lean years they had to sell their land just to keep going.

Jimmy Pitts of Pitts Lumber Company in Saluda said his family fell into that category. "My grandfather sold several thousand acres of timberland to Chesapeake during the 1930s," he said. "Chesapeake was the only place in the area that had any ready cash back in those days. My family appreciated the sale, too," Pitts added. "The cash from that land deal meant my family was able to pay back taxes and hold on to the rest of our land."

Slowly but surely Olsson gathered up a sizable amount of forest land for Chesapeake in and around the West Point area. As he obtained more and more timberland, he realized he would have to hire a forester to take care of it.

In 1934, Olsson hired a Yankee from Massachusetts by the name of Winslow "Windy" Gooch to be his first forester. Not only was Gooch the first forester at Chesapeake, he was also the first industrial forester hired in the state of Virginia. Gooch had all the credentials with a degree in forestry from the University of Maine on top of a master's degree in geology from the University of Michigan.

Gooch immediately set about establishing an authentic woodlands department with an office, small staff and all the proper "duds" that went into it. The little Chesapeake Corporation of Virginia was now ready to enter the big leagues with other paper companies across the nation with the beginnings of its very own forestry department.

Gooch's daughter, Mrs. Janet Redd of Virginia Beach, well remembered the excitement of her father and family coming

Winslow Gooch, Chesapeake's first "forester," was also the first industrial forester hired in Virginia (early to mid-1930's).

down to West Point, Virginia, at that time. The town of West Point was undeveloped at this time. First Street along the York River, which is filled with lovely waterfront houses today, was still mostly wild empty lots.

By 1935 the Gooch family had built a lovely home on the First Street waterfront next to the old Tyler Bland house. Although Gooch later left the Company and went on to work with the U. S. State Department and was involved with the World War II effort in France, he was able to kick off the very first official forestry work for the Company.

One of the very first important accomplishments for the new forestry department was public relations for the Company. Even in those early years, Chesapeake needed to present itself to the public as a caring company interested in conserving nature. Gooch was successful in pushing through the first seed tree law in the Virginia General Assembly which initiated the first steps for conservation in the state.

The new law required farmers and wood dealers to leave a certain number of cone-bearing pine trees during each cut. Gooch was mainly responsible for getting this legislation passed. He also introduced the concept of "selective cutting," a system where only a certain number of trees are taken, always making sure enough trees are left behind for reseeding and environmental purposes. In those days, this was an almost unheard of practice.

Known as Windy because he could talk an ear off a farmer before he was done with him, Gooch was relaxed, congenial, wore a crew cut, and had a good sense of humor. He was the perfect first forester for Chesapeake. He also started the first tree nursery for the Company. He did a lot to educate farmers on the Eastern Shore on how to "strip-cut" their woodlands so as to leave some trees standing as a wind break and buffer zone. This practice was needed to help conserve the ever eroding soil and as a seed source to reseed the newly harvested strips.

Janet Redd remembered her father's special dress whenever he was going into the woods for the day. "He wore leather putties [protective gaiters wound around the lower leg used by soldiers during World War I], a hat and high top boots," she said. In a day when foresters wore suits and ties to work, Gooch still managed to look every part a Virginia gentleman forester from Chesapeake, even while tramping through the woods.

Chesapeake's second forester, Captain Ed Tokarz, shown while serving in Germany in 1945.

Gooch left to work for the State Department in the 1940s but before leaving the company, he managed to hire Edmund Tokarz, a graduate from Virginia Tech, as the official "second forester" for Chesapeake. Gooch also hired Bud Johnson, the "third forester," a son of a Virginia Methodist minister from the western part of the state who had dual degrees in forestry and geology from Yale and Virginia Tech.

Tokarz soon left for military duty

with the war effort and with Gooch gone to the State Department, Johnson was left to take on the forestry duties. He hired a forester with a perfect name for the job, Forrest Patton, from the Virginia Extension Service, to assist him during those war years.

Johnson was also a congenial fellow, a flashy dresser, who always smoked a pipe, wore sports coats and bow ties. Like Gooch, he also fit right in with public relation duties and working for legislation to enhance Virginia forests. During his long years of service to the Company, he served several terms as Chairman of Virginia State Conservation and Economic Development Board. He was also honored as the Virginia Forester of the Year.

His wife, Florence Johnson, remembered their lives in West Point where their home was on Second Street, or what they called, "Hell's Half Acre," which was "one block from the water in two ways." He was a major contributor to Gooch's big drive to pass the Virginia Seed Tree Law.

Johnson, who was well known within his profession for taking at least a full minute to respond to any sort of stimuli, used to enjoy saying that while he was with Chesapeake, he "learned every woods, forest and country road in King William County."

But Johnson was not just a slow talker. He could be slow in other areas, too.

A favorite story that passed around the Woodlands office about Johnson was a time when he and Tom Tyler, a Company forester from Eastern Shore, spent a night in a cabin over on Eastern Shore before getting up the next day for a planned duck hunt. On the way out, Tyler inadvertently slammed the door on Johnson's hand. Since the door did not quite close properly, Tyler slammed it again two more times cussing up a storm as was his usual style.

Finally Johnson responded as politely as ever, with no extra measure of speed or urgency, which was his natural policy. "Please, Tom, don't slam the door again. It won't close because my hand's in it."

Another classic story starring Johnson and provided by Bob Kellison, Director of Forest Technology at Champion International, grew out of his love of ice cream. The 28 Howard Johnson flavors were his favorites and he was known to travel miles to partake of his special flavor, which was black cherry.

But even more than black cherry ice cream, he loved one of his own concoctions: black cherry ice cream garnished with the added delicacy of cooked lima beans. He soon became famous within Chesapeake for this dessert, which he ordered every time he had a chance.

Once Johnson, Claiborne Courtney and others from Chesapeake Woodlands were conducting business at the A. P. Hill Army Base in Virginia. When it came time for lunch, Courtney decided to have a little fun with his boss. After arriving at a restaurant, he took the waitress aside and told her they had a gentleman at their table who was going to be committed to a nearby insane asylum immediately after lunch. He added that she was not to be upset if he ordered anything strange but to just go ahead and bring him whatever he wanted.

The waitress kept her distance from Johnson all through the meal by speaking to him through Courtney. All went well until it was time to order dessert. Sure enough, Johnson instructed the lady to bring him cherry ice cream covered in cooked lima beans.

That did it. The lady bolted and spent the remainder of the meal safely ensconced behind the counter. Someone finally brought Johnson his dessert, but he complained to his group as they were leaving that he could not understand why the service had been so poor.

YALE UNIVERSITY
SCHOOL OF FORESTRY
NEW HAVEN · CONNECTICUT

January 14, 1944.

Mr. J. H. Johnson, Forester,
    Chesapeake Corp. of Va.,
    West Point, Va.

Dear Johnson:

Would be glad to have the chance to pass on the problem you outline, if you think the Company would consider me _unbiased._ I know personally that my opinions are always based on actual facts on the ground, but these opinions do not always coincide with high authorities.

As to charges - a _minimum_ of $300 — which for ten days is $30 per day, and far too low for any professional expert except a god dam forester. Thirty dollars per day for additional time, and time to include travel and writing report, plus any stenographic cost for which I pay 75¢ per hour. Plus expenses.

The important thing is, how soon? I am tied up with a big case right now involving Menomines Indians and want some time to work on it, but if I knew if and when the Company might want this done I would try to plan it to comply as nearly as possible.

Saw Steve Spurr yesterday. Quite a coincidence!

Sincerely,

H. H. Chapman

This letter and statement (opposite page) is from H. H. Chapman, a famed author of forestry textbooks, while performing consulting work for Chesapeake.

```
Chesapeake Corporation
   West Point, Va.                    to          H. H. Chapman
                                                  205 Prospect Street
February 15, 1944                                 New Haven, Conn.          Dr.
_____

To services and report

Expenses:                                                        $300.00

Jan. 22.  Coach fare, New Haven to Richmond, Va.
          Car fares
          Room, New York, required to catch a.m. train Sunday     10.03
          Tips                                                       .15
     23.  Two meals                                                 5.50
                                                                     .35
     31.  Railroad fare, Washington to Baltimore                    1.95
          Extra fare on Montrealler at New York, not
             provided in ticket                                      .97
          Supper
          Tip                                                        .99
          Carfare                                                   1.95
          Two telegrams at 60¢ each                                  .10
Feb. 4.   Stenographer on report                                    .10
             10 hours at 75¢ per hour                               1.20
          Postage                                                   7.50

                                                                    ___.30

                                                                 $331.09
```

Johnson had tremendous political punch in Virginia and his contacts over the years were a great help to the Company. Jack King, chief of Woodlands today, remembered that if anyone needed to talk to the governor of Virginia about a problem, all they had to do was tell Johnson and, "We'd get a call from the governor within the hour."

Charles Finley, past executive vice president of the Virginia Forestry Association, said he remembered many times Johnson included him on personal meetings with politicians at the Virginia State Legislature. Johnson was so laid back he often forgot to introduce Finley to the various legislators as they made their rounds. "I called it tag-along lobbying," Finley said with a laugh. "But many a time he picked up my lunch tab when I looked in my wallet and found I only had a few dollars," he added.

Gooch also brought on board a fellow Yankee with degrees from Ohio State University and the University of Michigan by the name of Forrest Patton. Patton was well known in Virginia and he had a reputation with all the farmers in the state for giving an absolute honest appraisal. Nobody would think of selling a piece of timber until Patton would personally walk the land and give his advice.

Over the years, Patton was well known at Chesapeake for a piece of advice he gave about the universal fear most people had about walking in the woods and the possible chance of stepping on a snake. "Remember," he would say to others with a smile, "if you don't look for snakes, you won't see any!"

One day as he was walking through the woods, a big black snake fell out of a tree and hit him directly on top of the head. Plenty of Chesapeake people teased him about not seeing the snake so it must not really have been on his head. He retorted with his ever present sense of wit, "If I had been listening for a snake that day, I certainly would have heard that fellow coming!"

Patton was nicknamed "Pappy" when the younger Dick Brake came into the territory and took over as chief forester. It was a nickname that stuck to Patton until he retired. Brake remembered Patton as a good forester who "took his job seriously but also enjoyed the good side of work."

Everywhere Patton went, he liked to hand out pencils with his name, his title, and the Company logo on them. It was surprising how far those pencils traveled. One Chesapeake employee reported he checked into a motel in North Carolina once and found a "Forrest Patton" pencil next to the telephone pad!

Bud Johnson and Claiborne Courtney on Chesapeake lands in the 1950's.

When Patton retired, he ordered a new batch of pencils that read "I'm a retired, tired forester." But Mrs. Patton said that shortly after the pencils arrived in the mail in 1993, her husband died. "Many who had known him called to ask me if they could please have one of Forrest's pencils as a special keepsake of an early Chesapeake Woodlands forester," she said.

Another big name in early Woodlands history was Randall "Red" Highland who arrived at Chesapeake in 1940 from West Virginia with a degree in civil engineering. He came on board as a surveyor hired by Cecil Woodward, who was then chief of Woodlands. Tom Harris, who became the head of Woodlands in 1968, remembered Highland as a "first class, fine,

Working on the form class tables are Forrest Patton (left) and Jim Girard, originator of the tables

41

Christian gentleman," an easy going fellow who had an honorable reputation and this meant a lot because to many wood dealers and farmers, he represented the Company. From the start, Chesapeake's Woodlands had the reputation of honest dealing and always being a company that was as good as its word. Highland was the major force behind this good reputation.

After Cecil Woodward's abrupt departure in 1946, Highland became the official head of Woodlands. Carl Olsson, the older brother of Sture Olsson, (who had been assistant to CEO McDonald) took over the Woodlands Division in 1952, the same year Sture Olsson became CEO of Chesapeake. Highland specialized in procurement of wood and Bud Johnson specialized in land management. It was a division of labor that worked but with some expected friction.

Dick Brake, who became chief forester and later headed up Delmarva, Chesapeake's real estate development subsidiary, remembered Highland was so respected and well thought of that nobody ever called him anything but "Mr. Highland." "Everyone else in the Company was known on a first name basis, even Sture Olsson," Brake said with a laugh. "But not Red Highland. He was always Mr. Highland."

Forrest Patton "cruising" timber in the 1960's and sighting in his hand compass near a large loblolly

Sture Olsson had a great respect for Highland. "He always said to new people, 'Stay out of the wood business if you're afraid of hard work!'" But Highland was not afraid of hard work nor were any of the staff he gathered together at Woodlands. He worked his people hard but at the same time he was well-liked, which is a rare trick for any manager.

Olsson went further in his description of Highland. "He was loved by all people in the wood business. Although he was not a 'graduate forester,' he did a great deal to improve the lives and professions of all the people he ran into who dealt in wood."

Carl Olsson was vice president of Woodlands but his years at Chesapeake were difficult years for the Woodlands Division because of his health problems which led to chronic absences. His staff remembered the stress in the office revolving around the fact that he was only able to come to the office once or twice a week. And then he never stayed much longer than to shuffle a few papers around on his desk.

Mary Ann Fetterolf, who came on board in 1959 as a young secretary for Woodlands, described those difficult years when Johnson and Highland were trying the best they could to run the Woodland's business on their own. It was not an easy task.

The day would start off with an absent Carl Olsson and a demanding CEO calling over from the mill on the other end of the ringing phone. "When is he coming in?" Sture Olsson

Red Highland (left), Carl Olsson (center) and Bud Johnson at the Eltham tract in 1958.

would shout over the telephone, as mad as a riled up rooster. "Where is he? What's going on over there? I need such and such papers! I expect official progress reports delivered on time from everyone and every department in this Company and that includes my brother over in Woodlands!" he would storm. Fetterolf said that after those calls everyone would work extra hard to do their best to keep things going.

"When Mr. Carl finally did show up," Fetterolf continued, "Mr. Sture would get him on the phone and there would be some terrible shouts that could shake off a door from its hinges." Finally the call would end. Highland and Johnson who were waiting outside the door would then go in, shut the door and each start talking about what had to be done that day, trying to get all important decisions made. Then Mr. Carl would get up and leave and the process would be repeated a few days later.

Mary Ann Fetterolf, secretary in the Woodlands office.

In spite of the dysfunctional system, work got done. Carl Olsson's staff loved him and they did all they could to keep things going smoothly and make sure all the work was done. The love for the Olsson family passed from father to both sons and that was how it was with the staff.

"Carl Olsson was a kind person," Fetterolf remembered. "He always brought me expensive gifts from his business travels. Once I remember he told me he hated some coral necklace I used to wear in the office. 'I hate that color!' he told me, then the next time he was away on business he brought me an expensive gold necklace from New York City. Once he even had shell earrings made especially for me in Key Largo."

The Chesapeake family was loyal to both of the brothers and loved them equally. "Yet, they were so different," Dick Cartwright, retired surveyor, remembered. "Sture was outgoing, you could hear him coming down the hall from a half mile away. But Carl was different. Still, you could never find a better man than Carl.

"One day Carl just walked up to me as I sat working at my desk and signed his name on my drawing board. I still have it, too," Cartwright added proudly.

But Mr. Carl had a fiery temper too. If he came into the office and found someone else had parked in his spot, directly in front of the door, there was trouble. Once he found someone parked there he would shout, "Who dared to park his car in MY parking place?" Then, "Get that damn car out of here!" This was a scenario that was repeated again and again when some unwary visitor made the dire mistake of parking in the wrong space!

Fetterolf knew firsthand how difficult it was for Highland and Johnson in those years. "They did the best they could to keep things going as well as possible." Even the big boss,

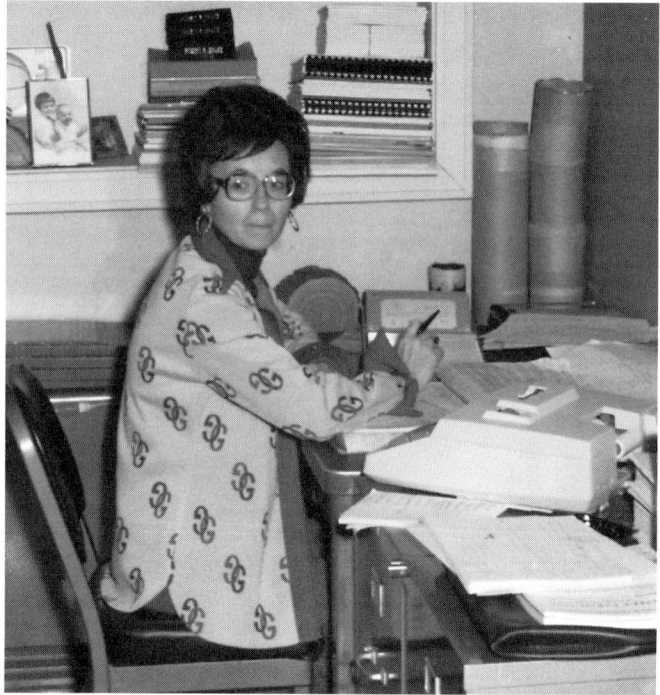

Mr. Sture, understood the friction Johnson and Highland had to work under and that it did not help their own relationship at work. "Bud and Red did not really get along too well," Sture Olsson recalled even years later, "but everyone understood the reasons why."

Until Ida Dawson another beloved and long serving secretary to Woodlands, came on board in 1974, Fetterolf was the only female working in Woodlands. "Those men were always great to work for," Fetterolf said with a laugh. "There was something about foresters. You couldn't help but love them."

Mrs. Redd remembered how her mother used to describe foresters, " 'Foresters are a special blend of science, warmth and intelligence,' my mother always said." Few professions can claim such an intricate incorporation of perfectly wonderful traits.

Ida Dawson, secretary in the Woodlands office, circa late 1960's.

Fetterolf might or might not have agreed with this definition. Surely she was in a position to know after all those years of observing up close the true nature of foresters. It hardly seems feasible that any of the foresters who passed through her direct eyesight over her 29 years of service to the Company could ever have guessed she would one day describe in a book what these men were like to work for. "But I was young," she chuckled, "I was running a one-woman office back then and it was always a lot of fun coming in to work each day."

Chapter 7

Ushering In New Times

Carl Olsson died suddenly in 1961, at just 45 years of age, and the several years following his death were filled with turmoil while the Company decided what step to take next. Red Highland was appointed manager of Woodlands while the Company decided what the next step should be.

It was clear to CEO Sture Olsson that the Company needed to modernize its Woodlands department in order to bring it up to the level of other companies in the highly competitive paper manufacturing world. This would mean expanding the forestry department with the best possible staff of managers that could be found in the industry. In addition, Woodlands would also have to develop a sound, long-term plan for replanting and managing pine trees on Company lands in order to meet the demanding needs for more and more pulpwood.

Woodlands' secretary Mary Ann Fetterolf remembered that the years between Carl Olsson's death and the arrival of Tom Harris in 1967 were particularly stressful with many changes in leadership. Mr. Overton D. Dennis, Chesapeake treasurer and member of the Board of Directors, was appointed vice president of Woodlands and he commuted to West Point from Richmond each day. Highland continued as Woodlands manager and soon thereafter Dick Brake took Bud Johnson's job as chief forester. At that time, Johnson was named director of public relations and chief lobbyist for the Woodlands Division.

Red Highland was now approaching retirement thus putting Woodlands in jeopardy of losing his many years of experience in bringing in the wood. Olsson knew, perhaps more than anyone else, that the loss of this very important link in the company had all the makings of a pending disaster.

This was the same year Sture Olsson, with the advice of other management, replaced Dennis as vice president of Woodlands and hired Tom Harris from the Albemarle Paper Company in Roanoke Rapids, North Carolina, to head up Woodlands. He they charged Harris with the task of completely modernizing the forestry department for Chesapeake. This was a timely decision in that Chesapeake Woodlands Division was lagging way behind the rest of the industry and needed new leadership to direct a game of fast catch up.

Tom Harris was just the man for the job. A native of Macon, North Carolina, Harris had a degree in forestry from North Carolina State University. He also had years of military service in the Army as a captain in the infantry that had landed him in Normandy during World War II just two weeks after the invasion. He was exactly the right man for the Company, the times and all the many challenges ahead. In addition, Harris had the right temperament and personality for the job, a crossbreed between a good-natured, pleasant, silky-eared spaniel that everyone liked and a hard-driving, sled-pulling, dedicated husky that would never take no for an answer.

One of the first problems that faced Harris was where to put his family. In the late 1960s, the town of West Point offered few amenities for upper management Chesapeake paper mill families.

Fortunately, some farsighted soul at the mill had earlier thought to buy five or six houses in Chesapeake's name so they would have homes for their future hires, if ever needed. Harris moved his family from North Carolina into one of them and went right to work.

Olsson had brought in his new Woodlands manager not only as Woodlands chief but also as a vice president of Chesapeake Corporation. With this title, Harris had the power and support he needed from Corporate management right from the beginning to lay the big, ambitious plans for the future.

"One of the major challenges we faced," Harris remembered, "was to develop the Woodlands Division into a top-of-the-line department that could stand up to the best in the nation." Harris knew his first step would be to draw up a plan.

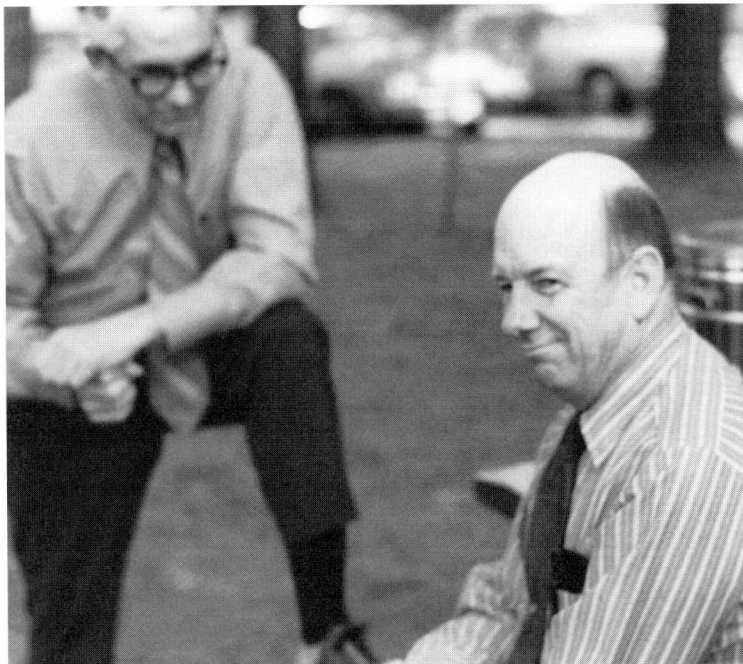

Tom Harris (right) consults with Paul Harper during a Company party.

After six months on the job, Harris prepared a corporate report early in 1968 that nailed down the problems Chesapeake faced. Based on that year's consumption of 623,000 cords of wood, 21 percent of the wood was brought in by truck, 33 percent by rail, 9 percent by barge, 21 percent of the wood came in chips from Pocomoke City, and 16 percent chips came by truck, it was easy to predict what future growth would mandate. Harris saw big trouble for the Company if drastic change was not begun immediately.

"The pine supply for this mill is critical due to overcut of pine in the immediate area and the low volume of company land," Harris stated in the very first point of his report. "The seriousness of the situation cannot be overemphasized."

Harris, no doubt, had every member of the Chesapeake Board of Directors held spellbound as he, point by point, explained every step they must make in order to rectify the drastic situation. He went on to recommend the following: Increase Company lands immediately at the rate of 20,000 acres a year, maximize utilization of hardwood in order to minimize pine consumption and cost, add four Company operated "concentration" yards to better control and regulate wood supply, mechanize pulpwood production because of the ever-diminishing labor supply, expand trucking by use of larger more efficient trucks, and continue site preparation and planting program on both Company and private lands at the rate of 17,000 acres per year.

After Harris made his report, there might have been a measure of silence around the room as every board member and upper level manager considered the huge jibe in the wind that such a plan would engender for the Company. It was as if Chesapeake was about to enter an entirely new era. It was to be the big leagues for Chesapeake from this point on.

"Bringing in the wood every single day for the booming mill was our number one concern," Harris said of his years at Chesapeake. "We brought in wood cut up in five-foot or eight-foot lengths in those days from all over the area on bobtail trucks, shorter trucks developed just for five-foot lengths of wood, [some trucks used were even the old Coca Cola trucks of yesteryear] a far cry from the big logger trucks filled with entire lengths of pine trees that come barreling into the West Point mill today," he added with a nostalgic look.

From the beginning, Harris was aware of the major flaw in Woodlands. It had been divided up into three divisions, procurement, land management and land acquisition. This made for an awkward situation that set each department up for constant inter-office squabbling and division.

Harris decided the solution was to split up all the Company land into six territories and have one man in each territory responsible for everything within his own area. This meant he would take care of all the procurement, land management, and land acquisition that needed to be done in his area. This one simple change in organization brought the entire Woodlands department together into one team.

Then Harris delegated specific duties to each manager, outlined job responsibilities and gave each individual full authority to get the tasks done. This was the first time anyone in Woodlands had been able to look at a piece of paper and know immediately what his job tasks were.

"Another change we brought about was in 1985," Harris said. "We sent each of our area managers to a management course at Pennsylvania State University. This was a great plan because it created business managers out of foresters and helped them talk the same language as the accountants who were slowly taking over the Company."

Even though Harris was drawing close to his own retirement by then, he had seen the handwriting on the wall. He knew his foresters needed to understand the business side of the mill operation if they were going to prevail.

But probably the most important change of all was developing the first real, long-range business plan for the Company that Woodlands ever had. Harris knew it took 35 years to plant, grow and harvest a fully mature pine tree. So he initiated a plan that would take Woodlands well into the next century with a plentiful supply of wood.

His comprehensive plan was threefold. First, they would continue actively buying forest lands for Chesapeake. Second, they would implement an aggressive reforestation plan. And third, they would shift a major portion of the wood procurement program from Virginia to North Carolina. "The irony is, in all the land we reforested, we never cut any of it all the years I was there because I considered those years our 'catch up' time. But I presume they are cutting it now," Harris said.

Harris was known for his honest, up-front leadership style. Jimmy Sears, then safety director, remembered that before Harris came on board Sears used to "pad" (What? Not Jimmy Sears!) his yearly budget requests "because you never got anything close to what you asked for."

"When Harris came, I asked for $675,000 for capital expenditures, which was exactly one third more than I needed," Sears remembered with a chuckle. "Then before I knew it, $675,000 on the dime had been approved. I hurried over to Tom's office and told him I really didn't need that much money and I never expected to get that much! Harris just laughed and told me that under him I was going to get whatever I asked for," Sears said, still greatly amused at the memory.

But something good came out of the extra money Sears received that year. "We built the

Four Woodlands chiefs gathered in 1988 are, from left, Jack King, Paul Harper, Tom Harris and Red Highland.

original chip mill at Keysville with those extra funds." And Keysville turned into a big moneymaker for Woodlands, so everyone was happy that Sears had gotten his extra money that year.

Outside of the daily demand of bringing in the wood to the mill, the greatest problem that Harris faced was the joint venture plywood plant with Champion International in Pocomoke City, Maryland. The business was called Chesapeake Bay Plywood Company and Dan Lewis, Chesapeake's representative on the joint venture board, was appointed to head it up. The new plant was nothing but trouble.

"It was a small plant," Harris explained, "and making a profit over there was difficult because the housing market was always in a state of fluctuation and this affected our profits." Finally in 1987, after expenses grew to be too high, the plant was closed. In spite of the frustration of running that sideline, Harris thought the overall experience was probably good for Chesapeake Corporation because, "It got the Company interested in expanding into other more profitable ventures."

Sture Olsson, however, summed up the demise of the plywood venture in more explicit terms. "We got involved with people who weren't our kind of people," he said. "We got into bed with better bean counters than we had and our bean counters never caught up with their bean counters and they beat us." It was good fortune, according to Olsson, that after

49

Harris arrived the bad venture with bean counting came to an abrupt end.

Harris said the Company also experimented with the wood treating industry and was very successful in developing certain wood products for deck and patio uses. When the Company eventually decided the venture did not fit with its long-range plan of core business, it pulled out of the wood treatment business all together. (More about this in future chapters.)

Harris was a beloved boss of the old school, with solid, traditional values. He was a fine southern gentleman. Whereas he always expected top performance from his people, he treated everyone in the Company in a fair, firm but friendly manner.

"You had to get up early in the morning to beat the boss into the office," one of his staff remembered. "I would rush out the door by 7 a.m. and find Harris already at his desk and reading." He had a slow to rise but "substantial" temper, too.

Jack King, present Woodlands chief, recalled how the men could see how the boss was feeling about something just by looking at the top of his receding hairline. "When his head turned red, we knew it meant to get quiet and back off!" King laughed.

Another forester remembered once someone drew a comical cartoon and pinned it up on the bulletin board where it stayed for many months. It depicted Harris sitting at his desk going over the rising costs and constant delays of wood shipments to the mill. His head was colored bright red and "sparking" just like a light bulb.

But perhaps the most important change that came with the arrival of Harris was salaries changed for the better. "We hardly made anything before Harris came," Eddie Bernoski, longtime Chesapeake employee, remembered. "Oh, I guess we made what the janitor made back in those days. But, with Harris our salaries went up considerably. And we all got real job titles too, even job description, with written goals, and regular promotions."

Even better, according to Bernoski, were the official area assignments that came in with Harris. "For the first time ever we knew where our areas were and we had specific duties in these areas. It really improved our work situations and how we felt about our jobs."

One of Harris's personal favorite memories of all the years spent as chief of Woodlands was a sunny spring day in May in 1972 when the Company celebrated the planting of its "100 millionth" seedling on the bank of the Mattaponi River at Sandy Point in King William county. This tree symbolized Woodlands' reforestation of over 130,000 acres of Company land in both Virginia and Maryland. The land chosen for the 100 millionth tree was the original site of a 17th century grant to William Frazier, whose descendants held the land until 1929 when Chesapeake purchased it. (It was once the family home of Amy DeFarge, the wife of Tau Crute.)

It was a major ceremony in Chesapeake history. On hand for the big day were Sture Olsson, chairman of the board, George Dean, Virginia state forester, and Edward Cliff, who had just retired as chief of the U. S. Forest Service.

"Well, we planted that seedling up right with all the proper fanfare," Harris said with a chuckle. "But darned if that 100 millionth tree didn't turn right around and die on us! I don't know how many times Chesapeake got in there and replanted that famous 100 millionth tree and, sure as anything, it died on us all over again. Finally we had to give up. We figured the soil was just too sandy to take the seedling. But since Chesapeake has owned that tract, we have planted and harvested three crops of pines." Harris laughed. "You would think after growing all that pine all that time on that tract of land, we could have started just one more seedling!"

Planting Chesapeake's 100 millionth seedling are, from left, Bud Johnson, Dick Brake, State Forester George Dean (planting tree), Tom Harris, Sture Olsson (with shovel), Chief Forester Ed Cliff of the U.S. Foresty Service, and Red Highland.

Probably the most famous remark that Harris ever made while chief of Woodlands was this quote that lives on to this day. "Hell will freeze over before I will ever hire any women foresters!" Harris had announced with a measure of heat to his staff. One can close ones eyes and imagine the North Carolinian saying it, maybe with a few red sparks shining off the top of his head.

When the first female foresters came on board, as they inevitably did in the 1980s, Dick Vasey, a forestry professor at Virginia Tech with a great sense of humor, immediately picked up his telephone and called Harris.

"Say, Tom, isn't it a little chilly over there in West Point?" Harris was the first to laugh when he heard the big joke was on him. Tom Harris was that kind of man.

Chapter 8

The Wood Dealers

No book about Chesapeake Woodlands would be complete without a chapter on one of the most interesting facets of the Company, their wood dealers. They certainly qualify under this category, and also as one of the most important ingredients of Chesapeake.

What a rare and interesting breed is the wood dealer. Shrewd, tough as the trees in the woods, feisty, hardworking, fun-loving, good-natured, and dedicated to getting the job done, the wood dealers always added a measure of color and excitement to the work day. They could sometimes be a hard drinking crowd, too. Woodlands chief Jack King remembered with a chuckle that when he first came to the Company, he was told if he wanted to do business with the wood dealers . . . he had better keep a bottle of whiskey in the glove compartment of his truck! But those days are long gone. To indicate just how much times have changed, anyone found with a bottle of whiskey on the job today would be dismissed immediately.

Chesapeake had a full network of wood dealers that was spread all over the state and other areas. These dealers were mainly the responsible parties for arranging all the big daily shipments of wood into the mill at West Point.

Jimmy Sears was having fun with the wood dealers when he created this artwork.

Although this book can not possibly touch on every wood dealer who had business dealings over the years with Chesapeake Woodlands, we want to capture at least some in print for posterity. As change has come about at Chesapeake, and other paper mills, the wood dealers have had to make many adjustments. But during their heyday, they were the most colorful lot of all.

There was probably no other wood dealer who served longer and worked harder for the Chesapeake Corporation than the John Vassar family in the Charlotte County, Virginia, area in the central part of the state. There have been four generations of Vassars who have been responsible for bringing in the wood for Chesapeake: John Vassar, Lealon Vassar, Lealon Morris Vassar, and Greg Vassar. No other Chesapeake family can make that claim in Virginia.

Lealon and his father, John, first started providing wood to Chesapeake

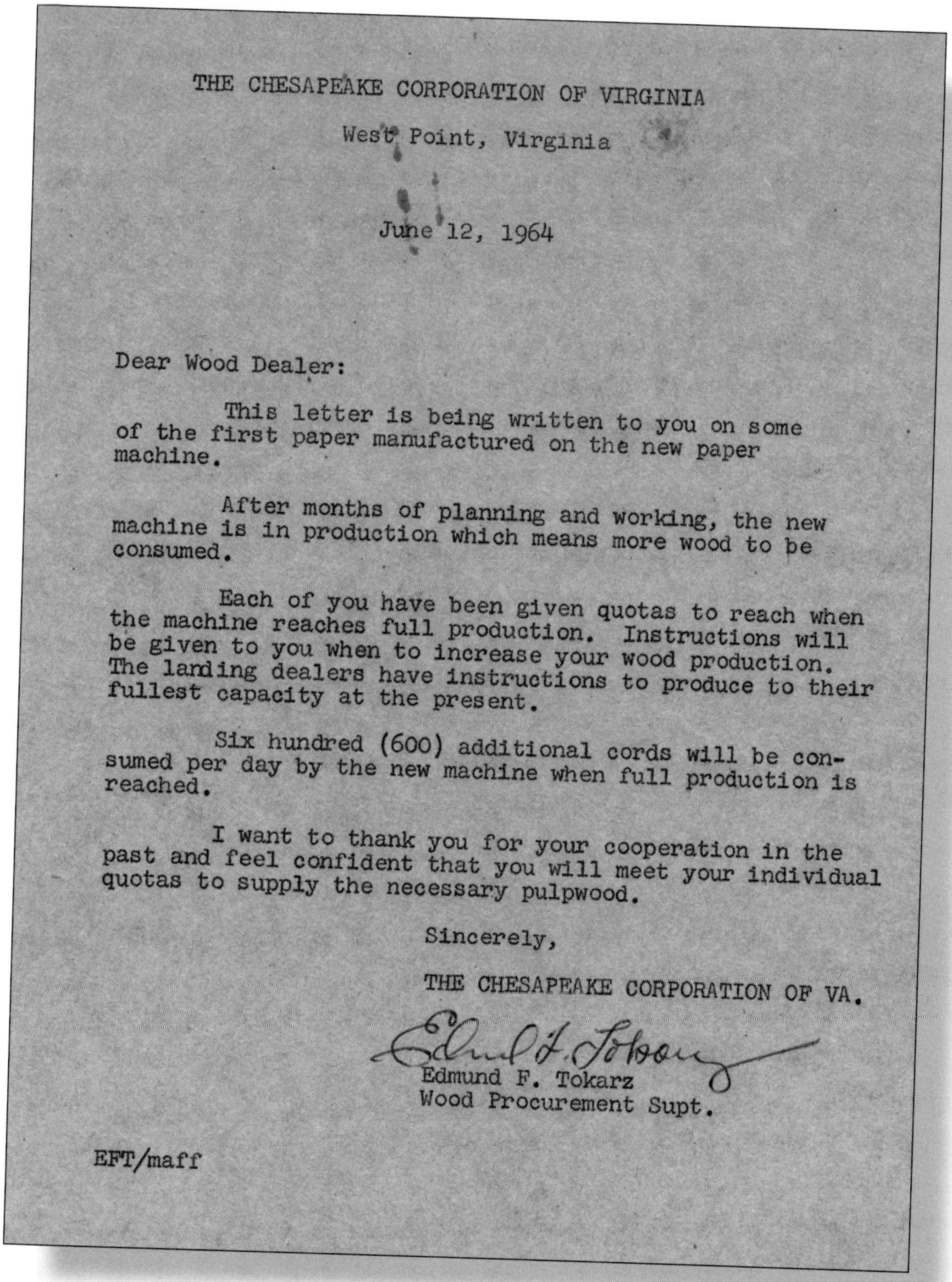

THE CHESAPEAKE CORPORATION OF VIRGINIA

West Point, Virginia

June 12, 1964

Dear Wood Dealer:

This letter is being written to you on some of the first paper manufactured on the new paper machine.

After months of planning and working, the new machine is in production which means more wood to be consumed.

Each of you have been given quotas to reach when the machine reaches full production. Instructions will be given to you when to increase your wood production. The landing dealers have instructions to produce to their fullest capacity at the present.

Six hundred (600) additional cords will be consumed per day by the new machine when full production is reached.

I want to thank you for your cooperation in the past and feel confident that you will meet your individual quotas to supply the necessary pulpwood.

Sincerely,

THE CHESAPEAKE CORPORATION OF VA.

Edmund F. Tokarz
Wood Procurement Supt.

EFT/maff

This 1964 Chesapeake letter to wood dealers was printed on brown kraft paper.

in 1946. Lealon was born in 1919 from hardworking Methodist stock in the Keysville-Lynchburg area and was first affiliated with Chesapeake when, in 1946, he saw a nice piece of woodlands for sale—over 10,000 acres that he thought could be purchased for back taxes.

Lealon Vassar called Chesapeake Corporation and talked to then chief Red Highland who sent Jimmy Sears, his right hand man, right out to look at the land. "Sears walked through the property and assured me the wood was good," Lealon said. "Even though no paper was ever actually signed with Chesapeake, I went ahead and bought the property for back taxes and knew I could count on the Company to buy the wood from me."

Lealon and John Vassar began harvesting the wood. In those days they worked their men in the woods with a single International truck and a two-man crosscut saw. Then real progress came to the business. "In 1949 we started using a two-man power saw," Lealon said with a

John L. Vassar (circa 1950).

laugh. In those early days, they loaded up the wood in boxcars and sent it over to West Point by rail.

Later, the closed boxcars of yesteryear were replaced with wood racks, the open train cars used today, which are easier to load and much more efficient. The new open cars allowed them to ship in at least 20 wood racks a week, each containing 18 to 20 cords of cut wood. (A cord is a quantity of cut wood equal to 128 cubic feet, in a stack measuring four by four by eight feet.)

Lealon said the fact that Jimmy Sears was able to guarantee him that Chesapeake would buy that much wood from him each week kept him going even through the lean years.

John Vassar had a stroke in 1954 and his son Lealon took over the business along with Lealon's son, Lealon Morris or "Little Lealon." Now even a fourth generation, John Vassar's great grandson Greg Vassar, is learning the wood dealing business from the ground up. "Greg is now running every piece of equipment we have in the woods and we use him as a back up man whenever and wherever we need him," Lealon said proudly. "He is also learning how to appraise timber," he added. "This skill takes years to develop."

The family worked hard over the years in the wood business and it was not always easy. "The biggest stress of all was with a full roster of workers, we were finally producing more wood than Chesapeake needed," Vassar remembered. "And keeping workers coming in to work every day and being able to pay them was our number one concern." Over the years the Vassar family kept buying woodlands and adding to their own land. Now they can har-

vest their own wood if they can not get any timber from neighboring farmers.

"Trust was always a very important part of this business," Vassar explained. "We knew we could trust Chesapeake and the farmers knew they could trust us." This was important because landowners did not always know how much acreage they had or how much timber was on it so they had to count on being able to trust the wood dealers.

Lealon Vassar said he appraised wood by walking the boundaries of a tract for an hour or so, as close to the boundaries as the farmer could tell him. "I could sort of just feel the size of the tract during that hour's walk," he said with a laugh. This is a real talent to be able to do this and it is not learned overnight.

Lealon G. Vassar (right) was among those demonstrating this two-man chain saw in 1957.

One day a farmer called him to appraise a tract and told him he had 100 acres. "Well, I knew I could do about 100 acres in an hour's walk around the lines. We were back in 35 minutes! So I told the man regardless of his survey, his tract was much less than 100 acres," Vassar said. Sure enough, when the farmer had his property resurveyed, he found Vassar had been right.

In 1982 a big change came about for the Vassar family. With Chesapeake's help, they invested in some big, expensive and state-of-the-art logging equipment to handle their work. They purchased two feller-bunchers, machinery that goes right into the woods and snips a full size tree off at its base. Skidders drag the tree to a deck and a loader then picks it up and lays it on the extra long trucks capable of carrying a full size tree to the woodyard. Consequently, the work getting wood out of the forest and to the woodyards was simplified. This, in turn, meant fewer employees and smaller payrolls. Such modern machinery changed the wood business for the Vassar family overnight and forever.

Lealon Vassar laughed when he recalled all the new troubles that faced wood dealers. "Insurance premiums, high financing, environmental concerns, accidents, personnel problems . . . but Chesapeake works with us and keeps us trained," he said. "And, we can call Jack King, present chief of Woodlands, at the West Point office and get help with any kind of problem.

"We always had that kind of personal and close relationship with the Company," Vassar said. "I remember once one of Daddy's paychecks was lost somehow for a load of wood and in those days that was a really serious thing. Daddy called Mr. Highland and he got right on it and before we knew it we had a new check in hand. Mr. Highland always said if we had any problem, any problem at all, to be sure and call him."

Vassar laughed a lot when he talked. "You know, we could always tell when things were bad over at the mill in West Point and they were short on wood. They would always come over to Keysville and get us out of bed so early in the morning even we early rising wood dealers noticed it! Then they would give us breakfast in a local diner and take us right into the woods.

"Over the years we noticed something about Chesapeake foresters," Vassar chuckled. "They always came the earliest in the morning and the foresters from Westvaco, an eastern U. S. paper company, and Georgia Pacific could never beat them!"

The Vassar family enjoyed many Chesapeake stories during their long years of business association. The question still remains which party was more shrewd.

"Once my Daddy complained to Mr. Harper back in West Point who was at the time managing Woodlands, that he needed a dollar raise," Vassar said. "Mr. Harper said he couldn't afford to give Daddy a raise unless it's 50 cents. Well, Daddy told Mr. Harper, if he couldn't afford but just 50 cents, then he didn't want any raise." Vassar laughed. "A couple of months later Mr. Harper cut my Daddy a dollar raise!"

Now the Vassar family is supplying St. Laurent, previously Chesapeake, with up to 500 cords of wood each week and has enjoyed in recent years big increases in price due to the increased competition for pulpwood in their immediate vicinity. Although other paper companies are in the area, Chesapeake still enjoys the reputation of buying the most wood. (This comment made in 1996 interview.)

"Chesapeake never lets us stay in the dark about economic situations and we appreciate that more than anything else," Vassar said. "With my big equipment and workers, we are fully invested for 40 hour work weeks even when we can't work because of weather conditions." Vassar said that the business knows both good and bad times. "One thing I know now when tight times come that I didn't know before; eventually the bad times pass and the good times will come back."

Lealon Vassar said he still loves to go personally into the woods and work on the skidder. "It's quiet there and there are no interruptions from the telephone or people bothering you. There is nothing like going to the woods at 6 a.m. and starting work." Vassar threw back his head and laughed at past memories and many good times. "And one more thing, in spite of all the changes, Chesapeake is the only company that still has foresters who will meet us at that hour, too!"

One of the most colorful wood dealers Chesapeake enjoyed over the years was Lynchburg, Virginia's very own, one and only, Bob Sales. Now there is a name that is known, respected and loved throughout the state.

Bob Sales was distinguished by his valor at the landing in Normandy during World War II. He was the only man in his all-Virginia platoon who survived the first day of the invasion. He later won a medal

This Vassar family photo taken during the 1990's includes "Papa," Lealon, Greg and Mathew.

for his outstanding service to his country. (see story in appendix)

While Bob Sales' valor and extraordinary service to his country in time of war were great, so is his personality. A man who was singled out by fate to be the sole survivor of his landing craft cannot possibly be a shy, reserved wallflower. And rest assured, he is not. Not by even the wildest stretch of words.

Bob Sales is fun. He likes to tell a good story and he is full of them. He loves life and it plainly shows with every word or laugh that he gleefully passes on to the rest of the world.

In his fine and glowing presence, one can't help but sit back and enjoy him. (This writer even received at the end of her interview with Sales a gift from this master salesman . . . a vial of her very own supply of genuine sand from Omaha beach! If anyone can top that, please let me know.)

Sales got his start in the wood business by going to see both Red Highland at Chesapeake and also, just in case Chesapeake did not pan out, the head man at the White Pulpwood Company. He ended up getting an order from Chesapeake. Woodlands agreed to take seven or eight carloads of wood a week. That was all it took. Sales was in the wood dealing business.

Sales remembered with great relish something his competitor back then said when he heard he was going in the wood business. It was F. T. Faulconer of Lynchburg, a local wood dealing powerhouse, who said, "That boy Bobby Sales will never do you no good, he's just too wild!"

And wild he was. No one could out party, out fun, outtalk, or out anything Bob Sales. With his personality and knack for getting people to do exactly what he wanted them to do, in no time at all, Sales built a successful wood dealing business.

Who could say no to a man like Sales? If any person tried to turn him down on an order of wood or a tract of land or even some dinky little favor, why, Sales would sit him down and have him in tears in a matter of minutes describing how he barely managed to survive the war and did he really want to turn down a genuine American hero in his time of need? Why, there's not a man or woman alive who could turn down Bob Sales, or if there is, I have not yet met the low-down scoundrel.

In those years, Chesapeake was growing rapidly and Sales remembered that by 1966 he was sending off 50 to 60 carloads of wood a week to West Point and making good money. The biggest problem Sales had to deal with was getting those empty train cars lined up so he could fill them up with wood. "I gave a lot of Virginia hams away over the years to railroad personnel to line up those cars," he said with a hero's wink.

In time the competition for those empty Southern Railroad wood racks got so hot, Sales used to pack a bottle of whiskey under the seat of his car and pour some out for the stationmaster. "Once a conductor and I were enjoying a drink of bourbon together so much he missed his own train! I had to carry him all the way to Monroe as fast as I could in my Olds convertible so we could catch up with the train and put the conductor back on it!" Sales said with a laugh.

His wood rail stops ran from Monroe, Arrington, Faber, Amelia, Camp Pickett, Chula, Dale City, Woodbridge and Winterham on its way to the West Point mill. Sales had wood going on the rail at every stop.

Sales remembered the lean times when the mill was overloaded with wood and they could not take enough from him. "So I called Mr. Highland to complain about the orders," Sales laughed. Highland told him to come on down to the mill and talk to Carl Olsson about the problem. "I had just bought a brand new Olds convertible so I hopped in the car and sped over to West Point," Sales said. "Then I got Carl Olsson in his office and told him I was going to the damn poorhouse with the puny wood orders he was giving me!"

Sales remembered that Carl Olsson stood up and looked out the window. There was his brand new Olds convertible sitting pretty as a princess right there in the Chesapeake parking lot. "Carl looked back at me and said, 'Well, Bob, as far as I can say it looks like you're the only damn wood dealer I know who's going to the poorhouse in a brand new Olds convertible!'"

Sales was known to be a man who was full of fun and Chesapeake employee, Jimmy Sears, remembered everyone at the mill loved going over to Lynchburg because they knew with Sales they would have a good time. "He was single and would party a lot," Sears said. "He also kept a real good expense account and would treat the boys to a good time!"

One day Jimmy Sears and a few others from the mill had to be in Lynchburg working over New Year's Day. They got to grumbling about having to work on a holiday until Bob Sales came by. Sales ended up taking them all out to dinner, then to watch the football bowl games over at his house.

Later that night they were all hungry so they went out to dinner again. Sears told Sales he did not have any money with him and asked him if he would please pick up the tab for the second time that day. Sales gladly paid the bill and put the double dinners on his expense account and sent it over to West Point.

When Highland looked over the account he picked up the telephone and called Sales on the double. "How in the devil could you and my staff eat two dinners in one night?" he shouted over the telephone. Sales laughed. "It wasn't easy, boss!" he answered.

Highland complained plenty about having to pick up the expenses, but under Sales' fast smooth-talking charm he finally agreed to pay the bill. Only Bob Sales could have gotten away with it. After all, he was the only survivor at the landing in Normandy and a genuine American hero. Highland knew if he did not agree to pay, Sales would have him feeling terrible over his great sacrifices in the late war.

One day Sales remembered Carl Olsson had come over to ride in a big parade honoring the annual pulpwood queen through the streets of Lynchburg. "It was a really hot day," Sales said "Our convertible was in between a troop of cowboys on horseback and a marching band that hit more bad notes than good. Halfway through the parade Carl turned to me and said, 'Hey, Sales, have you got any liquor?' I answered, 'Why sure I do, Carl, why as you talk, you're sittin' on two pints!'"

The two men pulled out the bottle and each took a good, swift swig, under cover, of course. From that point on, the parade went a whole lot nicer for both men. The occasional horse droppings off the forward bow or the off-key blaring tubas from the rear were not so bad after all.

Sales spent 40 years bringing in the wood to the Company. Over those years he had plenty of expense account meals. "I used to have every waitress in Lynchburg trained to give special service when I had what I called a 'live one,'" Sales laughed.

"Chesapeake had a bunch of good people, Sears, Henley Everett, a lot more, and every spring they threw a big party over at West Point for all the wood dealers with lavish food and drink," Sales remembered. "Now those days are over and a lot of the personal touch we used to enjoy is gone."

Sales felt he was every bit as much a Chesapeake employee as the men back at the mill. "They get pensions and I don't," he added, ever ready to play the violin about his suffering during the you-know-what. "But Chesapeake was a great company and they did a lot for all of us," he said.

Robert Sales during World War II.

Over the years Sales remembered Chesapeake would take a loan out for a worker so he could afford to buy his first truck and also helped the wood dealers buy the big equipment. "This was a big favor because none of us could have gotten money from a bank for the loans."

Sales remembered Roy Hunt, who was his procurement field man in the Lynchburg area for many years, and also Henley Everett, Dave Ware, Carlton Edwards and Charlie Evelyn. "Highland once invited me to come to work for him at the mill," Sales said. "It was a major mistake on my part that I didn't take that job," he added with a sigh.

A measure of sadness filled the air as I prepared for my exit from our interview. I saw the handwriting on the wall. Bob Sales' earlier rendition of war stories had left me dabbing my eyes with a hanky. But Sales was going to work me over one more time for a little extra sympathy about his never going to work directly for Chesapeake.

I got up and left, pocketing my own private supply of genuine sand from the landing on Normandy. I made a fast exit, too, because I could see the old pro wood dealer from Lynchburg was about to squeeze one last tear out of me.

Ralph Clements in the Lynchburg, Virginia, area was another long time wood dealer for Chesapeake. A past lefty pitcher for the St. Louis Cardinals (He won 12 and lost one.), he eventually quit baseball and entered the insurance business in Richmond, Virginia. (Even while providing wood to Chesapeake he continued with Nationwide in a Lynchburg office and finally retired in 1994.)

Clements first got into the wood business when he married Miss Mattie Emerson from the Dillwyn area of Buckingham County. The two were married on the pitcher's mound at the ballpark in Durham, North Carolina.

Mattie's father, John Emerson, was in the wood business and Ralph started working for his father-in-law. "I got my $40 a week salary and did pretty well and when Mr. Emerson died of cancer in 1958, Mrs. Emerson offered me a 50-50 deal if I helped her run the woodyard."

It was a deal Clements could not turn down, so he went to work still holding on to his insurance business on the side.

Mixing up the insurance business and the wood business was an interesting experience for Clements. "It was suit and tie with the insurance people in town and then I was in overalls with the racket of buzz saws slicing up sticks and sawdust everywhere with Chesapeake," he said.

He remembered he would often be sitting in a posh office working with insurance problems and the telephone would ring and a voice in West Point would tell him, "Ralph, somebody's stealing our wood over on Buzzard's Roost Tract. Would you go over there for us and check it out?"

The wood business made Ralph a lot of money over the years and he sold about 75 percent of his wood supply to Chesapeake even though he never had a formal contract with the mill. He figured that at his peak, he was sending between 60 and 70 carloads a week into West Point. The train stopped right at his woodyard to load up and made stops along the line at Howardsville, Scottsville, Columbia, Wingina, and Cohassette and loaded up with wood from his subcontractors. He paid them after Chesapeake paid him.

The wood was loaded by hand and as it traveled on to West Point, an interesting phenomenon would take place. The wood would all "shake down" and the cars would be less full than when they were packed down the line.

Chesapeake did the measuring by stick, a system they still use today, which has a measuring man actually walking along the wood racks holding up a special measuring stick and marking the wood already measured with a piece of chalk to prevent re-measuring the same wood. When the wood was measured at the mill, a paycheck would be issued and mailed off to Clements.

Clements remembered getting his first financial break in life from a local fellow by the name of G. M. Davis who told him about a 40-acre tract of land he could buy for just $600 that had at least $1000 dollars worth of marketable wood on it. Ralph made the deal and never forgot the good deed Davis had done for him to help him get a start in the wood business.

Over the years Clements has made many other good deals in the acquisition of valuable land. "Still, it's almost a miracle today if you can get your money out of a 25-year-old stand of pine," Clements said. "You almost have to 'hook' someone to make a profit—that is, find a deal somewhere, somehow, that's worth more than you have to pay for it." This hook system has been a tried and true way to quick wealth long used by the wood dealers.

In the past a lot of timberland has been bought up at almost dirt cheap prices, sometimes merely for payment of back taxes. Much of the huge acreages of Chesapeake woodlands was bought at unbelievably low prices by today's standards. "But most people are using consulting foresters now before they sell," Clements said. "They now know exactly what their land is worth before they make the deal."

Clements retired from both the insurance business and the woodyard in 1994. But his son, Johnny Clements, is running the woodyard now along with woodyards at Cumberland (in addition, he had operated another woodyard at Cartersville) to make it a third generation of wood dealers for the Emerson and Clements family.

One of Clements' favorite memories was Chesapeake's Jimmy Sears bringing a young Chesapeake intern down to his woodyard in Dillwyn one summer while the young man was off from classes at the University of Virginia. Clements sized him up. "He was a cocky, intelligent fellow, destined to do big things in the Company," Clements thought.

"I never forgot that fellow's name either," Clements said with a laugh. "That young intern was Carter Fox, who later became CEO at Chesapeake."

Trying to pin down wood dealer Randolph Fogg to the exact area of his family's four generations of service to Chesapeake is like trying to grab a bubble. So where do the Foggs do business? Easy. "Anywhere there grows a tree," said 84-year-old and now retired Randolph Fogg of Oak Grove near Colonial Beach, Virginia, in Westmoreland County.

Randolph Fogg really comes from wood dealing stock. His father, Samuel Fogg, began providing wood to the West Point mill in 1920 by floating logs down the Mattaponi River out of Maracossic Creek near Aylett where they were picked up by a Company scow. It was probably the good Chesapeake ship "Ethel" that Randolph saw inching her way in and out of the creek. There was no better craft than the "Ethel" for getting into shallow water and picking up the sticks.

He even remembered at a very young age going down to the mill at West Point with his father to that back office on the river side to talk with Mr. Clyde Goldman who was vice president then and in charge of sales. "I used to see Mr. Elis Olsson sitting in the office next door." In those days, for a young boy getting a glimpse of Elis Olsson was almost like seeing God. One can imagine the big-eyed look through the office door and the swift intake of breath at the vision. The young boy never forgot the image of the famous paper-making Swede sitting at his desk tending to his work.

Samuel Fogg also had a sawmill and in 1946 his sons, Randolph and Randolph's younger brother, Manley, bought their father's sawmill for a "very reasonable price." Randolph decided against his dad's advice not to go to college. "I'd end up behind a horse in a field with or without a college degree," he told his father. There was nothing for the father to do but let his son do as he pleased. That was how Randolph was, headstrong and a man who followed his own inner voice, a sign of a very good future wood dealer, indeed.

Randolph only had an 11th grade education and he swears to this day he only learned to multiply and add and never to divide or subtract. Another sign of a very good future wood dealer, indeed.

He remembered once during the depression telling his dad he was going to quit the wood business because wood was one of the last quick cash resources available to the natives and it seemed everyone was out in the woods chopping down trees. "In ten years there won't be a pine tree anywhere in King and Queen County to cut," he told his father. Samuel looked at his son and smiled. "Why, son, that's what they all said back in the days when I was a barefoot kid."

Randolph was not convinced. So his father explained why there would always be trees to cut. "Look, son, the trees are growing seven days a week and 24 hours a day and we're only cutting them down five days a week and eight hours a day!"

Still, Randolph eventually decided to move over to the Northern Neck above the Rappahannock River and settle in the Colonial Beach area. "I had to move north because everywhere I went in King and Queen County I could hear a darn sawmill running," Fogg explained, a rather discouraging sound for someone who wanted to start his own wood business.

He began to organize forces on the Northern Neck across the Rappahannock River. His two sons, Bob and Ricky, are now in charge of both the wood dealing business and the chip production business which they took over in 1976 and 1982 respectively and probably also at a "very reasonable price."

Fogg used trucks, barges, and even trains out of Milford to move his wood to the mill. Like other dealers, he sometimes had more wood than the mill could take, which caused problems.

The Company provided a lot of help in the areas of financing and good advice. Forrest Patton, a Chesapeake forester assigned to that area, was one of these advisors. "Forrest knew more about timber than any man who ever came through this town," Randolph said. Fogg remembered Patton's outspoken ways and how the man would always call a spade a spade, no matter what. "He would go back to the mill and tell them something or other they were doing was just plain stupid," Fogg laughed at the memory. How many foresters today would think of telling management something they were doing was just plain stupid?

"We loved him," Fogg said. "He really went to bat for his people." Fogg also liked the way

From left are father and son wood dealers Bob, Randolph and Ricky Fogg.

62

whenever that man from Chesapeake would run into some brush in the woods he would simply "take out his machete and start chopping his way through it when any other man would have gone around it." That was Forrest Patton. Nothing held him back, certainly not bushes.

Forester Walter Zingelmann eventually took Patton's position as advisor to the Fogg family. "I thought a lot of Zingelmann, too," Fogg said. "Paul Harper, then Woodland's chief, sent him to me when I asked for help and Walter really helped me work out my business problems."

Zingelmann was especially helpful at mealtime. "My wife used to say he always showed up at my house at lunchtime with something he needed to tell me," he laughed. "My wife always served him up lunch just like us and she loved it, too, because in those days there wasn't a restaurant in the area and she knew she was the chief cook!"

Fogg is famous in the wood dealing business for being the first to use a chipper in the woods, thereby revolutionizing the way wood moved from the forest to the mill. "I never could talk Mr. Highland into using a chipper," Fogg laughed. "But when Tom Harris came in as head of Woodlands, he gave me the go-ahead."

Fogg bought the ninth chipper ever manufactured by Morbark Machinery out of Winn, Michigan, in 1969. "I still have her, too, out in my garage and I'm never going to let her go." The new equipment cost $123,000 and Chesapeake gave him $100,000 in an interest-free loan to help him purchase the machine.

Fogg experienced good times and bad times just like all the other Chesapeake wood dealers. Once he was at a stockholders' meeting with Sture Olsson and he complained to some of the other wood dealers he was losing $1000 a month in his business. Later Olsson came right over to him like a fly to a lump of sugar.

"I understand you're losing money, James R., [Olsson always called Randolph Fogg "James R."] and I want to know why you're losing money," Olsson said, ever the sort to get right to the point.

"I guess I'm just dumb or something," Fogg told the king of West Point. But when a wood dealer tells you such a thing as he is dumb, watch out, it might be wise to fold your cards and run.

"Have you any idea why you're losing money, James R.?" Olsson persisted, never the sort to watch out or fold his cards and run.

Fogg finally admitted to Olsson he had figured out he needed to buy more equipment and then he could utilize his workers better and then turn his losses into profits.

"Well, then," Olsson said to Fogg, "I guess then you're not so dumb after all, James R., but plenty damn smart!"

Fogg had a natural ability to deal with his workers. He knew alcohol was absolutely lethal in the woods and devised a special way to filter out anyone who had been drinking from ever picking up a chainsaw.

"In a bunch of men you'll always have some 'sour krauts,' " Fogg said, using a special name he coined for someone who had recently imbibed. "So I would go around and sniff everyone's breath before I sent them into the woods. Whenever I found someone who was holding his breath when I walked by, I would poke him in the ribs real fast with my injured arm and he would have to take a breath!"

A serious chain saw accident, in which one of his workers fell in the snow on top of his running chain saw and went unattended, turned Fogg into a careful man when it came to whom he sent into the woods. "That leg got infected and ended up costing me $42,000 before we were finished!"

Fogg, too, experienced some accidents in the wood business. The worst was in the 1930s as he was crossing a one-lane bridge with a full load of pulpwood. He met head on with a bootlegger in a model A Ford fully loaded with hooch. The Ford ran right up on the truck and Fogg's right arm was almost severed.

"Lucky for me old Dr. A. W. Lewis lived down at the foot of the hill and heard the crash and came out running with his black bag or I would have bled to death right there on the spot," Fogg said. "Five doctors worked on me at the hospital and I could hear them discussing what to do with my arm. Three of them wanted to amputate at the shoulder but the other two said no." The two doctors prevailed and his arm was saved.

Chesapeake went on helping the Fogg family right into the third generation when Randolph Fogg's son Bob came out of college with an engineering degree and, after a stint with the Virginia Department of Highways, entered the wood business. "I didn't know a thing about forestry," Bob Fogg said, "so Chesapeake forester Jim Vadas took me under his wing and I went everywhere with him for an entire month." Vadas taught Bob so much about forestry that Bob always said, "I got my forestry degree in just 30 days of study from the Vadas School of Forestry!"

The family claimed its fourth generation when Bob's son entered into the family business. He has since moved elsewhere and now there are no members of this generation presently working in the wood business.

The Fogg family served Chesapeake well over the years and although the chipping business has run its course, the family has done well. The secret? Fogg owns 2500 acres of sweet timberland.

Fogg smiled that special wood dealer's smile when he mentioned all the land he owns. "You see, my daddy used to buy timberland for Chesapeake back in the days they did not have a woodlands department or foresters working to acquire Company land. Then I came along," Fogg laughed. "I bought land for myself!" he added, proving that Sture Olsson had the one and only James R. tabbed absolutely right when he said he was not so dumb after all!

Avery Owens was another longtime, cherished Chesapeake pulpwood dealer from the Eastern Shore who, through the years, provided help and wood to the West Point mill. Although he died before we were able to interview him for this book (just one week before the jet charter flew me over to Salisbury, Maryland, to meet with him), his son, Sam Owens, agreed to meet us to share his father's story.

The Fogg Brothers operation includes this chain-flail debarker leading to a chipper and then to a truck.

A knuckle boom loader feeds the chain-flail delimber/debarker and chipper in this view of the Fogg Brothers operation.

"Dad started supplying Chesapeake with wood in 1962 and was perfectly suited to do this because he had been a timber buyer before he met up with Chesapeake. I guess buying timber came naturally to him because my grandfather, Samuel Owens, had a sawmill. In addition to providing wood to Chesapeake, Dad managed a sailmaking business and also a hardware store up until 1975 when he retired from everything."

During his many years supplying pulpwood to the mill, Avery Owens managed a supply network of over 15 independent wood producers. Anyone who has ever managed anything knows this was a tougher task than the actual work of cutting and transporting the wood back to West Point.

"Many of these independent workers were unskilled workers from the deep south, Alabama, Mississippi or Georgia," Sam Owens remembered. "Dad would locate the tract of land he wanted harvested and take the crews in, but he had no direct power to hire or fire these men so he had great difficulty getting them to put in a good day's work.

"Dad said his major concern was getting the crew to cut within proper boundaries," Owens said. Some of those boundaries were very difficult to see with all the trees. Sometimes Owens had to work with very sketchy plats or archaic surveys that were based on a creek that might have changed course or a special rock that could have long ago been picked up and moved or an oak tree that might have died 50 years before.

Another problem in those early years was that the crews worked by hand and the work was very difficult and sometimes done in the worst weather conditions. If there is anything worse in this world than cutting timber in 100 degree temperature with every bug in the woods swarming around one's neck, it cutting those same trees in winter when the temperature is around 10 degrees and the snow is up to the top of one's boots.

When the high-tech equipment finally came into use on the Eastern Shore, it was not much help to Owens because the work crews did not have the skills to use the new equipment. There was no way to take an ax man who may not have even been licensed to drive a car and turn him over to him a skidder, bulldozer or chipper.

Owens said his dad referred to the years when the new high-tech equipment came in as the "crossover years." This was when modern, expensive equipment, sometimes costing a half million dollars or more to purchase, operated by highly trained workmen replaced the unskilled work crews of yesterday.

But the high-tech equipment caused a new set of problems. The cost of the new equipment was exorbitant and if Owens wanted to use the machinery, he had to sign a loan for it and make huge payments each month on the new equipment. Then he had to hire highly skilled workers and pay them much higher salaries. "On top of that we had to do on-the-job supervision ourselves, hire and fire our own people and even settle claims ourselves," Sam Owens reported. Before long it was not profitable in time or expense to be a pulpwood dealer.

But Sam Owens did say his father made money. "Dad made money, not from pulpwood sales but from finding out who had good timberland available for sale and at good prices and then being smart enough to buy it right then and there."

Sam Owens said his father worked long and hard providing up to 800 units of wood a week to the mill. (Each unit is 7300 pounds or one-and-one-fourth standard cords.) Owens remembered that his dad mentioned there could be a seven to 10-day lag between sending the wood to the mill and getting the check back in hand. Since his father had to meet the

The champion chain sawyer during the Pulpwood Festival at Amherst in April, 1955. Pictured are wood dealers, Mr. White, in the hat (left), and Bob Sales, to right of sawyer, in the white shirt.

payroll each week himself, this sometimes posed a problem.

Avery Owens saw shortwood pulpwood logging come into style in which the logs were cut into five-foot lengths for shipping. After that came the process used now where the entire length of the tree is sent to the mill. He also saw OSHA come into being and suddenly his father had to follow a long list of new safety rules and provide his men with hard hats and heavy boots and carry insurance to pay for any injuries.

Sam Owens laughed remembering one of his father's stories about trying to meet all the new government rules. "The men didn't like the new equipment they had to wear in the woods. The hard hats were hot in summer and the boots with their steel toe reinforcements were heavy and ungainly. Dad decided one Christmas to buy every one of his men a new pair of safety shoes for a gift. The next week he asked his crews how they liked their new boots," Sam said.

"They're no good," one disgruntled worker spoke up as the rest stared gloomily at the boss. It looked like the presents had been a big flop.

"Why aren't they any good?" asked Owens in bewilderment.

"When we stood next to the fire to warm our feet, the metal burned our toes!" was the reply.

Chapter 9

In the Woods

The forest is a beautiful place, a temple of nature, with trees standing in reverence under the sky, birds singing a chorus of jubilation, and the sun coursing through a web of green leaves from the heavens above, casting intricate patterns of light and shadow on the earth below. All who have walked through the woods at varying times during the day or year have experienced the beauty, have felt the closeness to nature, and have been left with a spiritual imprint of great awe and respect for this great green world.

Whether it be in winter at the first light of dawn when human breath smokes like a chimney in the frigid air, frost laces the bare branches of trees and a stick underfoot cracks like a report from a gun or in early summer with the new green leaves sprouting from limbs, insect, tree frog, bird, and squirrel chirping, chattering and humming like instruments warming up for a symphony, with the ever stark shadow of a hawk circling overhead with an eye that misses no movement in the trees below the forest is a world of its own.

There is an eternal call to the wild outdoors. It is no wonder many men and women choose to work in the woods as close to nature as possible, shunning all contact with the busy office of ringing telephones, blinking computers, demanding customers, and pressing needs of the boss. Some of these men and women for many years found their living working for Chesapeake.

This does not mean they lived completely carefree lives working in the woods. Those who know the outdoors are well acquainted with the natural irritants and downright aggravations of life in the woods. No one knows that more than Adam Geron of West Point, Virginia, who spent 38 years with the Company starting in 1956 when he was hired by Red Highland in a starting position in Woodlands as axman.

"I had just graduated from West Point High School," Geron remembered, "when I went down to Chesapeake to apply for a job and Mr. Highland came out of his office and saw me sitting on the bench with my lunch bag in my lap." Highland asked Geron, who came from a long line of Chesapeake employees, if he liked to work outdoors and when Geron said yes, Highland answered with a simple, "Fine, you're hired."

That was the way Chesapeake hired in those days. It was a Company that knew everyone in the area and liked to hire members of families who had proven through past performance that they were stable, loyal, and hardworking employees.

"Then Mr. Highland asked me what was in that bag and I told him my lunch and he laughed," Geron said all these years later. "I did not know, you see, that the work crews in those days always ate lunch together at noon in local restaurants." In spite of the passing of all those years, Geron still remembered feeling a bit embarrassed about having a lowly lunch bag in his lap.

Geron went right to work as an axman on the West Point survey work crew. One of his favorite memories of those "six wonderful years" he spent working in that first crew was of one day when he was over in New Kent County on a Chesapeake tract cutting a line out. "All five of us were spread out pretty far apart," Geron said, "when I hit a hollow tree with my ax. That tree had a honey bee nest in its base and the bees attacked! I turned and ran for my life!"

John Garrett, a forest technician, in a pine stand during the 1950's.

Geron laughed at the memory of youth and getting his first work experience on a job. "Only trouble was there was no place to run except back down the line because the forest was so dense I couldn't leave our trail," Geron explained.

"The other men saw me coming with those bees swarming at my head and they shouted, 'Don't come toward us, Adam! Turn around! Run the other way!' but I just came running

A typical woods operation in pulpwood (circa 1940's).

and I ran and ran and then they had to turn and start running too, shouting all the while for me to turn around.

"By the time it was all over I had been stung 28 times, eight times in my head. I remember lying down half dead on the ground and seeing one last bee who had lost his stinger still humping me on my arm for all he was worth. I remember flicking him off with my finger with the last drop of energy left in my body," Geron said. "Boy, later did I have a gigantic headache!" But Adam was still enjoying memories of when he was young and the days were filled with not only hard work in the woods but also good comradeship.

Back in those days the work crews worked hard and played hard and Geron remembered them as his best years at Chesapeake. "Our crew was headed up by Dick Cartwright, head of surveying, Dallas Wyatt was our instrument man and our axmen were Gabriel Fostek, Carroll Dixon and myself," Geron said. "We were close. When we worked over in Eastern Shore, we used to party at the Wagon Wheel bar after hours or play poker in our motel rooms at night."

Geron started taking some drafting courses by correspondence after work and before long he had left the woods and was doing drafting work in the office. Eventually he was doing title searches as part of the land purchasing and surveying program and also planning the daily work schedules for the two work crews at West Point and Eastern Shore.

He also helped run surveys in those days. Some of that land had never had a real survey done. "The paint crew, Buster Davis and Bobby Wilson, would come through after we ran

71

A Chesapeake family. From left are James H. Geron, 31 years with Chesapeake; George H. Geron, 35 years; Adam F. Geron, 38 years; Robert M. Geron, 35 years, and father Adam C. Geron (seated), 42 years.

a survey line and paint the trees along the boundary."

Some of those old homesteads had wells on the site that had been covered with a century's worth of vines, and Geron remembered that several times he almost fell into one of those old wells. "They were a constant and very real danger in the woods because you couldn't see them as the tops had rotted away and vegetation hid them completely," Geron said.

"Dick Cartwright, head of surveying, was my favorite person to work for and we had a lot of fun," Geron said. "When anyone in the Company needed anything done from a huge job to a little problem, they would always call surveying and know they could get help from us."

What was it like to spend a day in the woods? Geron remembered how hot it could be and how thirsty and hungry you could get after a long day of grueling work. "Sometimes we would come upon a bunch of wild grapes and we would stop work and pick those vines clean and swallow the sweet grapes, seeds and all," Geron said. "I still remember how good those grapes tasted.

"Once in the Northern Neck up on the Potomac River, we came to a cove on the river and we started digging up oysters and washing them off in the water. We cracked them open and slipped those raw oysters down our throats just like jelly," Geron remembered. "Oysters never tasted better to me than that summer day up on the Potomac."

72

Geron remembered the extreme thirst in the woods, too, especially on hot days. "We might be out of water supplies and suddenly come up on a stream in the woods and we would lie down on the ground and drink that cold water and never once were we sick. You sure couldn't do that today," he added.

"Then we were chased by a lot of farm animals too, especially if we had to cross a pasture or open field and some pig, cow or bull would come after us and we had to run for the closest fence and get out just in the nick of time," Geron said with a laugh.

Those were the days. Young, employed, laughing, male, ax in hand and free as the wind in the woods. Memories of work in the woods never got any better than that.

A member of Adam Geron's work crew was Carroll Dixon who started with Chesapeake Woodlands in 1964. Originally hired to join the surveying team, he was soon driving every sort of heavy equipment used to clear land and prepare sites for new seedlings.

Dixon well remembered the heavy equipment and especially the bulldozers winding up for action in the early morning woods preparing for site preparation. Sometimes it was so cold the men could see their breath. They would hover around a barrel fire to enjoy a last cup of coffee out of a thermos before climbing onto the machines.

This was a time when the woods had seen the last of the logging machinery that had so recently stripped and harvested the pine. The woods was now mostly cleared and ready for Dixon's crew except for patches of trees left in specific areas for nesting or soil conservation. The rest of the land was filled with a tangle of scrap shrub, stumps, broken trunks, and left-over wood debris.

It was now time for the work crews to come in and clean up the mess, prepare the good earth and plant a whole new crop of lovely pine that would grow to full glory in the next 20 to 30 years. It was an act done every day by Chesapeake Woodlands and other forest products companies. It is a process very similar to a farmer's plowing over last year's field of corn husks and readying the land for the new seeds that would soon bear new corn. For trees are planted and harvested similarly to many other crops.

"First came the noisy drum choppers," Dixon said, and they would bumble and stumble along mashing down the odd assortment of debris to prepare

This seedling planting crew in 1940 included Mr. Kurdziel (left), R. Bujack (right) and two unidentified workers.

A bulldozer pulls a "chopper" over harvested land prior to a prescribed burn.

for fire. By midmorning the bare ground was covered with flattened vegetation and leftover debris that would be set on fire and burned under the watchful eye of the Woodlands crew.

Another way the crews cleared debris in the early days was to pull a long chain between two bulldozers. That would knock over the leftover vegetation too small to be harvested. This was a difficult task for the bulldozer operators because they would lose sight of each other in the heavy brush and one driver would inevitably get way ahead of the other.

"Dick Brake was always thinking up new schemes to improve how we cleared the land and planted new trees," Dixon laughed. "And Sharon Miller did, too. He was always reading articles in some trade magazine about some better idea on how to do the work. But like most theories, we found it difficult to apply the better idea to our actual work."

Dixon remembered the new seedlings went in by late winter or early spring, which is the perfect time to start a new pine crop. In the 1960s, Woodlands was still planting their new seedlings by hand. The crews walked together in lines, stopped at certain intervals, opened the earth with a quick slice of a sharp tool that looked like a square knife, and inserted a new tree in the slot. With a simple boot heel applied to the earth, the slot was closed tightly around the new seedling. Tomorrow's glorious pine tree was off to the races.

By the mid 1960s, Dixon remembered, they had started trial planting with a planting machine that cut a trench in the earth. The machine operator then inserted the tiny trees in the open ditch and packing wheels closed it. But the crews found this exercise highly frustrating because of the rough terrain or many existing trees or stumps that interfered with the path of the trench. It was one rough ride for the person operating the planting machine. "It was more trouble than it was worth," Dixon added.

The work crews eventually began a system in which they worked as a team with as

A new pine seedling growing after a "seed tree" cut (circa 1940's).

many as six tractors chopping side by side until the entire site was prepared. "We had a lot of fun working together," Dixon said, "especially when the first man on the line would turn up a bee's nest with his blade and the rest of the men would jump off their tractors and start running!

"The bees, mostly yellow jackets, were big problems in the woods," Dixon remembered. "Once you stirred up a nest, the bees would attack your tractor and stay with it, swarming as mad as raging bulls for several hours. There was nothing you could do but go off somewhere as far away as possible and do some other work until the bees finally gave up and took off."

Dixon said the original tractors had canopies but no side screens and therefore offered little protection to the driver. "Some of us would spread mosquito nets around our tractors in an attempt to protect ourselves. Once I was stung 12 times from one hive but I was all right. In all those years I only witnessed one man who ended up in the hospital after a bee attack!"

A metal sign used for marking a seed tree in the 1940's.

Dixon said a big change that came about over the years was the new equipment. It now comes with enclosed cabs with heat and air conditioning to keep workers comfortable in the woods no matter what the weather. Enclosed cabs also afford protection from bees and insects and even better, protection from falling snakes. Dixon remembered an occasional snake would drop down off a tree, which happened to him on one unforgettable day. Snakes falling out of the sky are no longer a problem with an enclosed cab.

"We would travel all over the Chesapeake system, work all week Monday through Friday, on one location, staying in motels each night and then returning to West Point on weekends," Dixon said of his work schedule. "We went to Keysville, west of Richmond, the Eastern Shore, and the Northern Neck areas and worked on all the Company land. We worked a hard schedule but we had fun, too, and we were all very close to each other."

Dixon's favorite memory of his many years of work at Woodlands was an old stray dog they named "Junkie" that had attached himself to the forestry garage, the shop and office for the site preparation group. "He would always be there in the morning to meet us at the gate and we all enjoyed feeding him. On the weekends, Sonny Gresham, or one of the rest of us, would go by the garage and feed him. We could not bear him waiting until Monday before he got another meal."

Dixon laughed remembering how Junkie was nice to everyone except two people, the man from Virginia Power Company and Vic Kaczmarski, a fellow Woodlands' employee! "No way would Junkie let those two men in the gate!

"One day Junkie came down with heartworms and we all collected money to take him to the vet for treatment. Junkie got well again and all was fine until the new supervisor, Mike Harbin, came along and banned the poor dog from the grounds."

This was not the end for old Junkie, however. One of the men on the crews, Tommy Cobb, had taken to riding old Junkie up on the tractor with him as he went about his work. "Tommy was so attached to the dog he couldn't stand the thought of dumping Junkie off at the pound," Dixon said. "So he took Junkie home with him and the dog lived happily ever after."

Dixon looked off in space for a moment as if considering his last words on the subject. "Funny how an old worthless stray dog like Junkie could end up meaning so much to so many people," Dixon said.

Pennsylvania-born Vic Kaczmarski came to Woodlands in 1960 via his first job with Chesapeake, which was working the tough shifts over at the paper mill. "I loved the outdoors so I was glad to leave the mill and join Woodlands during a time they were expanding personnel, back when Bud Johnson was the Chief Forester in the Department."

Kaczmarski worked three or four years taking a correspondence course in forestry and was soon working in land management, planting trees and doing site preparation work. "In those early years we were still planting our trees by hand and never thought the machinery worked as well at placing seedlings as the men did."

King William County was Kaczmarski's territory where he served as an area technician. "I did everything there was to do in the woods, ran bulldozers and motor graders, planted trees, evaluated timberland, bought timber, you name it, I did it."

When Kaczmarski came on board, there were three official timber buyers for the Company, Bobby Owens over on the Eastern Shore, Ralph Turner in the Piedmont area, and Tau Crute in the local area.

Kaczmarski saw himself as the center of a wheel of problems, trying to keep all the spokes oiled. The needs of the landowners, the Company and the private logger crews all had to be met one way or another. This was not always easy to do.

"I saw myself in somewhat of a protector role. First, I had to see to it 'my' loggers were taken care of properly as loggers don't know much about money matters," Kaczmarski said with a laugh. "I tried to buy enough timber in 'my' territory so that the loggers always had wood to cut. Then I had to pay a fair price to the landowners to make sure they got a good deal, and at the same time, try to get the right price for the Company so the deal would benefit them, too.

"I bought land from both the rich and the poor. I found I could work with all kinds of people on a one-on-one basis. And my wood estimates always paid out just right, too," he said with a sly smile.

Kaczmarski remembered how slow the title searches were in those years and how long it took to record a deed in the old clerk's office at the Courthouse. "The company was buying land so fast over in King William County that sometimes we bought and cleared timber before the title searches could be done or the deeds recorded in the clerk's office."

Kaczmarksi also remembered the big discrepancies in the old and new surveys. "Sometimes there would be as much as a 400-foot difference in the old plat I was holding in my hand and the actual boundaries that the new survey crews had discovered."

Such discrepancies caused a great deal of problems for Dick Cartwright in surveying, but discrepancies went with the territory. Old surveys mentioned lines with creeks that have shifted directions or dried up, old sheds that were gone, a 100-year-old oak tree that had disappeared, but new surveys are run on exact measurements and steel stakes inserted in the ground.

One truth in the wood business is that here has probably never been a woodland survey without some question from someone on at least part of the boundary line.

But when Chesapeake found a discrepancy that showed a landowner owned less than he had thought, "those farmers could sure get hot," Kaczmarski remembered with a laugh.

Romantics like to think there is nothing finer than a lovely stroll in the woods. In real-

Wood being removed by crane from a "Bobtail" truck and loaded onto a barge (circa mid-1950's).

ity, the weather conditions often ensure that it is either too hot or too cold to really enjoy walking in the woods. And when walking in the woods is uncomfortable, working in the woods is even more so.

Eddie Bernoski of West Point, Virginia, a longtime employee of Woodlands, remembered just how hot it was working in the woods for Chesapeake, especially during the summer months of July, August and September when they did most of their site preparation burns.

"You have to remember that in 90 degree weather while doing a burn in the woods, it was really 110 degrees!" Bernoski said with a smile. "We did most of our burns in the summer months to clear out all the dead wood and prepare the soil for planting in the late winter and early spring."

Burn-overs are a part of good forestry practices when working with pine. Pine calls for a good burn-over in order to control competing vegetation. Nature provides regular "good burns" by lightning strikes, but most burns today are planned and started by man. Unless lightning happens to strike at the exact spot where Chesapeake wants a burn, and at the right time of year, the crews have to spark their own fires.

The Chesapeake crowd must have automatically sensed just how much I liked to hear snake stories because it seemed during my interviews, I heard a lot of them. Almost immediately Bernoski launched into one of his "close encounters with a snake."

Once he and Chesapeake forester, Wayne Beasley, were working on site prep in Prince George County above Richmond. Suddenly Beasley shouted to him a sentence Bernoski will never forget. "Hey, Eddie! You just stepped on a copperhead!"

"Of course, I didn't believe him but, sure enough, when I looked down at my feet I saw this huge copperhead stretched out on the forest floor just like a hose," Bernoski said. The two men were deep in the woods that day and much too far away from civilization to get emergency treatment for a poisonous snake bite. "If that baby had been coiled and ready to strike, I wouldn't be here today telling you this story!" Bernoski laughed.

Bernoski's story illustrated a constant problem for Chesapeake workers. How to get out of the woods and to emergency medical help quickly if such service was needed. Unfortunately, much of the work that needed to be done was not at the periphery of the forest, but deep in the woods and far away from help.

Bernoski also remembered that once he and Beasley had cut open a big black snake they had just killed in the woods because they thought he looked "mighty strange." "Out popped three baby turkeys!" Bernoski said.

The insects were a constant problem for those who worked in the woods. This writer remembered touring a piece of Delmarva Properties, the real estate subsidiary of Woodlands, with George Geron and stopping to look at a stand of young trees that had been grafted.

Suddenly we were surrounded by a cloud of mayflies that seemed to come at us out of nowhere. The swarm was so thick Geron had to pull out a can of insect repellent and spray all around us just so we could get back to the truck.

Bernoski said bugs were a constant bother, especially the biting gnats of the Tidewater area, which he believed could pester a man more than any other insect. "Once I got a gnat caught buzzing in my ear when I was three miles deep in the woods," Bernoski said. "That damn little gnat sounded like a B29 bomber in there and I almost went nuts until I got to my truck and headed to a doctor. The nurse poured a glass of water in my ear, the bug drowned, and I lived," Bernoski said with a chuckle.

But bugs were big trouble for all the people who worked in the woods. On that particular day one little gnat had the capacity to stop Bernoski's work in the woods for several hours. That is just one more version of David meeting up with Goliath. But this time, David was a gnat and Goliath was a Chesapeake man busy in the forest, bringing in the wood.

William Goode of neighboring Gloucester County, Virginia, spent 44 years working for Chesapeake Woodlands. During that time he spent plenty of time in the woods. He first started working for the Company in the Marine Department, which was for many years a part of Woodlands. In earlier days, the various woodyards were scattered up and down the rivers and creeks, and Woodlands relied on their fleet of ships to bring in the wood.

This forest soil training session in 1965 included, from left, Bud Johnson, Ralph Turner, Dixon Jones, noted soil scientist Dr. Coile, Bill Carrigan, Eddie Bernoski, Bennett Vinson and Dick Brake.

"We would either be bringing in the wood by tug on the barges or we would stay in West Point repairing the docks in front of the paper mill," Goode remembered. But Goode kept getting bumped into other jobs by all the returning vets from the World War II effort who had been promised that their jobs would be waiting for them when they came home. "Once, when I was displaced by a vet, I even had to spend nine months over at the paper mill. I really hated that," Goode added with a shy smile. He is a man who only wanted to work out-doors.

About that time Chesapeake's first official aviator, an ex-navy pilot named Jimmy Sears, called Goode up to take him on his first airplane ride over to the Eastern Shore. "It was before the Company owned any planes so Jimmy had rented a little two-seater and we flew right across the bay. I remember Jimmy carried a box of red dye and he told me, 'Don't worry if we go down, Bill, we can mark the exact spot with this red dye!'" Goode laughed at the memory of this trip and how they finally landed over on the Shore in a bumpy field, hopped out of the plane and did an inventory of the wood.

"When it was time to return, wouldn't you know it but the weather had turned really bad," Goode continued. "Sears knew I was filled with apprehension. So to take my mind off my concern, he had me put my foot on the airplane brakes while he stood at the front of the plane and whipped the rickety prop. It finally started and we flew home in pouring down rain, rolled in at West Point, and pushed her back in the hangar," Goode said.

That's when Sears told Goode that the airplane they had just flown in to Eastern Shore and back had just been fixed after a recent crash! "I'm just glad Jimmy didn't tell me that until we got back home!" Goode laughed.

Finally Dick Cartwright over in surveying snagged Goode for some surveying work in 1956. He stayed in surveying until he eventually went into site preparation where he stayed until he retired in the early 1990s.

"I came into the Company in 1946 at a time when we still were using the old two-man chain saw." Goode remembered well the very first red and orange Mercury chain saws that came in at the end of the 1940s. "They were hell to start and after we finally got them going, they would shake so much they would shake to pieces!" But the men soon learned that if they kept putting the screws back into the chain saws, they could manage to keep them going.

Otho Eubank from the maintenance department soon took Goode under his wing and taught him everything he knew about the heavy equipment in Woodlands. Goode spent bad weather days when they could not get into the woods working on the engines back at the shop. Goode enjoyed this and what he was learning would eventually come to good use.

In 1967, Goode became site maintenance supervisor of a team that operated four or five big bulldozers, two medium bulldozers, three small John Deere 450 tractors, a motor grader and dump trucks. "We had 18 people and three mechanics on our team but as time went on, our forces dwindled because someone would retire or leave and they were never replaced."

When Eubank retired and Goode took over the maintenance of equipment in the late 1960s, he suddenly had the EPA and OSHA to deal with. "By then, Jimmy Sears, aviator, had turned safety director.

"One day he came over to the shop with a long list of things we needed to do to satisfy the new government regulations," Goode said. It turned out Woodlands had to take out oil tanks, put in septic tanks and make a lot of other changes to meet the new requirements. They went right to work on the projects until they met every new regulation right down to the last detail.

The EPA and OSHA kept a close check on what was going on at Chesapeake, and Goode saw a lot of changes over the years. "We would have to take a special course on working with hazardous fuels or some other subject or provide physicals for our workers every so often. Then we would have them check our underground fuel tanks to make sure there were no leaks," Goode said. "And if we ever planned to move a tank, we had to notify EPA and they would send a man over from Norfolk to supervise us and take soil samples."

Goode remembered when Dick Brake supervised his department. "Dick always supported me, backed me all the time, and under him our department produced more than ever." In fact it was Dick Brake and Red Highland who had made him a supervisor. "Those were the days in the Company when, if they saw you could do the job, regardless of your education, they gave you a chance to do it," Goode said.

As the years went by, Goode noticed a gradual change of attitude in some of the new workers. "So many of the new people didn't seem to care much about their jobs or even want to work." Another change that came about was the workers had to be qualified and licensed to operate only certain equipment instead of merely hopping on any piece of equipment and going to work the way the crews did in earlier times.

Eventually, with EPA, OSHA and other requirements constantly hitting on the Company, it simply became too expensive and too much trouble for the Woodlands department to keep up the heavy equipment and crews. It was much more cost efficient to hire out the work with private contractors.

Also, it was more cost efficient for the Company to use herbicides and prescribed fire as a means of site preparation in lieu of operating the costly heavy equipment. As soon as Goode retired in 1990, Woodlands closed up the old shop, sold all its heavy equipment, and early-retired the rest of the men.

Dick Cartwright, the famous Chesapeake surveyor whose name probably comes up more than anyone else's in this book, came on board in Woodlands in 1951, hired by Red Highland to fill a vacancy in Clarence Hildebrand's surveying crew.

Cartwright really wanted to be a forester for the Company, as he was originally trained at Virginia Tech. But since the job opportunity at Chesapeake was in surveying, he went ahead and took the surveying job. A short time later, Cartwright decided to take the Virginia state exam in surveying. He passed. From that point on he was too valuable to Woodlands as a surveyor to ever let him change over to his original profession of forestry.

When Hildebrand eventually retired as head surveyor, Cartwright was just the man for the position. A favorite story told at Chesapeake was when Cartwright passed his surveying exam and got his license, he went back to Highland and asked for a pay raise. "Why should you get a raise?" Highland responded. "You aren't doing anything now with your license that you weren't doing before without your license!"

But Cartwright had a mind of his own and after almost 10 years with the Company, left in 1960 to set up his own private consulting business. He could not help but notice most of his work in his new firm centered on surveying Chesapeake land.

Cartwright learned like everyone else who starts his own business that it is not easy. It is a sad fact in today's world that nothing but trouble and complications await the entrepreneur. To illustrate this point, Dick Brake shared a story about how his friend Cartwright ran into some "misunderstanding" with the IRS.

As the story went, Cartwright received notice that he owed a certain amount of money he was supposed to have deducted and paid for his employees. Cartwright called up the IRS steaming mad. He told the IRS in no uncertain terms he didn't owe any money to anybody. The IRS answered a bit stiffly that the money he was referring to was not HIS money but his EMPLOYEES' money. Cartwright got so mad he hung up on the IRS agent!

A short time later he got to thinking about what he had just done. He trotted on down to the office of his lawyer, Nelson Sutton of West Point, to ask for some advice. When his lawyer heard he had just hung up on the IRS, he took swift action. He reached for his check-book and wrote a check for the exact amount the IRS said Cartwright owed. "Here, take this check and get up to the Richmond IRS office just as fast as you can and apologize, too, or you'll go to J-A-I-L!" he told his enterprising client.

Cartwright finally decided the expense and hassle of running his own business was not worth the effort. Since it looked as if he was going to survey Chesapeake land whether he was in business for himself or working for the Company, he returned to Woodlands in 1969 as chief surveyor. Cartwright spent 20 more years heading up the surveying department and proudly claimed he "was the only forester in Woodlands that had been hired twice by Red Highland."

In the early years, Cartwright's surveying office was in the basement of the original, red brick office building that stood next to the paper mill. The two basement rooms were officially known at that time as the "Procurement, Survey and Forestry Department," quite a mouthful for just a few humble rooms below the main floor.

"Mr. Olsson had his office upstairs," Cartwright remembered. "During those first years when I worked in Woodlands, nobody in the Company really knew what I was doing although I was working surveys every day, because no formal records were ever kept of the work that was done in our office."

Cartwright described Highland's workplace as very laid-back and informal. "It was a casual office; he never asked and we never told." All this changed, however, when Tom Harris arrived in 1967 as chief of Woodlands and Cartwright returned to the company in 1969.

"When Harris came, it was an entirely different situation and one with which I was much happier," Cartwright said with his foxy smile, the kind of smile one might expect a surveyor to wear who was just about to redraw someone's boundary lines. "Harris was really good to work with. If you did a good job on a survey, he knew all about it," the surveyor added. "For the first time we were setting work goals and filing reports. We had a plan and we knew each day exactly what we were doing at all times."

Cartwright supervised three surveying crews, two in the West Point area and one based on the Eastern Shore during a time when the Company was acquiring new land at an extremely fast pace. The surveying schedule was busy and sometimes done at almost a frantic pace.

"At one time," Cartwright said, "we had up to 375,000 acres of Company timberland. Much of the new land we had purchased had never had a formal survey done on it and quite a bit of it had unclear boundaries of which nobody had any actual records." Such a situation could be looked upon as either a surveyor's dream come true or his worst possible nightmare.

Cartwright was a perfectionist, a detailer, and he really cared about getting his surveys right. Thus, he initiated a systematic program of surveying Company land "from scratch," which meant doing surveys that took the search all the way back to the very first record of land ownership.

Because of Cartwright's insistence that even the smallest details were correct on his surveys, his men on the work crews had a lot of fun with him. It is a time-honored dream and eternal entertainment, enjoyed by workers the world over, to work for a boss who insists that every little thing be perfect. His men used every opportunity to turn his unmitigated dedication to perfection into an adventure of pure fun.

"Once I had a crew over on Eastern Shore and we were walking the lines along a piece

The surveying department in 1988 included, from left (standing), Arthur Richardson, instrument man; Alex Fostek, computer draftsman; Dick Cartwright, chief of surveying; Jerry Wilson, computer draftsman; Bobby Wilson, party chief; (kneeling) Francis Fostek, chainman; Charles Kerns, surveyor, and Adam Geron, computer draftsman.

of Company property that had a huge marsh on it," Cartwright remembered. "I knew there was supposed to be about a half mile of marsh between high ground and a certain stone marker but I couldn't locate the stone." Cartwright would not quit work for the day until he located that stone.

He tried to get his crew to help him locate the stone but ended up having to do the job himself while the men watched from higher ground. It is possible some of the crew were assembled with smiles on their faces ready to enjoy the show.

"I walked 50 feet out into the marsh and I looked down and saw I was up to my ankles in mud," the surveyor said. "The men were starting to snicker from their vantage point but I kept on going. Pretty soon I was up to my knees in mud. But I kept going. Before I knew it, I was up to my waist in that mire and now the men were laughing out loud. 'The heck with that darn stone!' I shouted back to my men. 'Get me outta here!'"

Cartwright remembered hard work in tough conditions . . . heat, briars, bugs, marshland, and even snakes. "I ran into everything, deer, raccoon, fox, I even killed a rattlesnake once in the mountains." He always wore heavy boots in the woods because a surveyor looks for line chops, three marks on a tree, as he walks and doesn't necessarily watch where he puts his feet. This can be extremely dangerous in the woods.

"Cartwright had a lot of kid in him," his crew member Carroll Dixon remembered. "We had a lot of fun working for him, too. We worked hard by day but we would party at night.

"One night we stayed out until 5 a.m. drinking and having a good time," Dixon said. "The next morning the boss was knocking on our door at 7 a.m. as usual, ready for us to go to work. We looked out the window and it had snowed and I remember when we got to the woods the snow was piled up knee deep. We grumbled about having to work in such bad weather. But Cartwright never let us off. He told us, 'You can stay out late as long as you like, but you have to get up at 7 a.m., no matter what!'"

Sometimes the chief surveyor had to deal with an occasional drinker who thought he would come to work drunk and go into the woods to work. "One man started to come to work drunk so I took him aside and told him I didn't care if he was drunk at home but he couldn't be drunk at Chesapeake and keep his job. The good news was the man stopped drinking and his wife even called me up to thank me!" Cartwright said with a laugh.

Occasionally Cartwright was involved in a difference of opinion with a farmer over his boundary line and the Company's line. Cartwright did not always like the decisions the Company made on such disagreements. It seemed to him that whenever anyone disagreed with one of his boundaries and called the Company to complain about it, Chesapeake would always give in and make him go back and change the boundary according to the farmer's wishes.

"I heard plenty of, 'Well, my grandpappy said his property line was over there', or 'My mama swore we owned all the land between here and that old red barn,'" Cartwright remembered, "Well, his grandpappy or mama was wrong. But we tried to keep everyone happy." Sometimes Cartwright felt the farmers got the best of Chesapeake. A favorite story at Woodlands was about once when Cartwright and his surveying crew were over on the Eastern Shore working on a new piece of timberland. It was the first time ever the chief surveyor had worked a piece of land on the Shore. As he plowed right into the woods eager to set the right example for his crew, the men told him to be careful of snakes. "I turned around and laughed and told the boys there weren't any snakes in these woods," Cartwright said.

"Later that day the boys called me over and presented me with a brand new carpetbag someone had purchased. I took the bag and thanked the boys very much for the nice gift. Then I opened the bag, let out a shout, dropped the bag and ran! The boys had given me a nice big copperhead found on the new Company tract! That day I learned not to push the boys so hard while in snake country," he said with a chuckle.

Longtime employee and West Point native Bobby Wilson spent 43 years working as a surveyor under Dick Cartwright and others. The last 25 of those years he served as party chief of Woodlands' surveying crew.

Originally hired by Red Highland in 1951, Wilson said he thought during his tenure with the Company he must have walked every line of every tract of Chesapeake land in Virginia and never once saw a rattlesnake. Over the years, Wilson said, the Surveying Department was led by Highland, Hildebrand, and Cartwright and is now managed by Charlie Kerns of Gloucester, Virginia.

"We did a lot of work over on Eastern Shore," Wilson remembered. "We would fly over on Monday and come back on Friday. We also spent lots of time in North Carolina. We did everything there was to do including working in the seed orchards." The crews used machetes to clear the brush and over the years Wilson saw many accidents in the woods in which a flashing blade hit an arm or a leg that happened to be in the way.

Wilson, who never once had a sick day in 46 years, remembered measuring the wood on their racks every Friday. This was a task that sometimes fell on the survey crews along with the regular measuring crew. The job was done with a special 6-foot stick, and entailed making a mark on the wood with a piece of chalk to make sure they did not measure the same wood twice.

During transportation into the mill the wood would "settle in" the rack which caused some problems between the Company and the wood dealers over the pay that was owed. Wilson measured wood for 20 years for the Company. "All the wood is weighed now right at the landing, which is more exact," Wilson said.

Over the years, Wilson fought many forest fires, too. "The biggest fire we ever had to contain was in the Holly Fork area in New Kent County, Virginia," Wilson said. "We fought that fire for several days before we managed to put it out and it burned up a lot of Chesapeake acreage, too."

Wilson even worked some at the mill, which was unusual for Woodlands people. "Once we had a strike and the salaried people were assigned to the paper machines to keep the mill going," Wilson remembered. "In order to keep the mill open, each manager had to work two shifts. As I remember, the union settled the strike real quick."

Working for surveyor Dick Cartwright was a lot of fun, according to Wilson, because "Cartwright liked to have a good time. Once at a firemen's convention in Fredericksburg we were up all night partying but, sure enough, at 7 a.m. there was Cartwright knocking on our door telling us it was time to go to work!" Wilson laughed. "But if you want to know the truth of the matter, the boss was as slow as we were that day!"

"But Cartwright had a temper," Wilson remembered. "He would get all riled up about some landowner giving us trouble over a boundary line. Tom Harris would have to take him into his office and settle him down."

Still, Wilson thought his boss had good reason at times to be angry. What really "ticked him off" was occasionally a farmer would make an unreasonable demand on a boundary dispute. According to Wilson and others in the surveying crew, it seemed like Chesapeake always ended up giving the land in question to the farmer. It was enough to irritate anyone, let alone a surveyor who knew where the correct property lines really were!

After Tom Harris took over Woodlands, he began to perceive vast woodlands planted in perfectly formed loblolly pines or "super trees" growing in bountiful reserve on Company lands which would be ever ready for future use. But how to perfect what Mother Nature had produced and was already growing in plentiful amounts on most Chesapeake acreage?

Claiborne Courtney was just the man to develop the super trees. After a stint in the Navy, he came on board with Chesapeake in 1949, hired by Bud Johnson. Courtney was eventually given the job of creating a perfect tree. The first step, Courtney remembered, was to identify exactly what Chesapeake considered a perfect tree to be.

It could be that in his search to create a perfect tree for Chesapeake, Courtney had to become God. This would be quite a feat, even for a man as able as Courtney. Fortunately, he had a great sense of humor as he related his work in a way that no employee could ever forget. "When I was in the U. S. Navy I never dreamed I would end up pimpin' for a bunch of trees," Courtney always said.

This was how "pimpin' Chesapeake style" was done.

In the field during 1982 near a Chesapeake boundary marker are, from left, Bobby Wilson, Francis Fostek and Arthur Richardson Jr.

The Woodlands crew began walking through the woods in search of nature's super trees. What were the traits that Chesapeake wanted? The list included trees that had extraordinary height, best possible volume, a short crown with straight limbs, a relatively high wood specific gravity, good taper or form, disease resistance and a plentiful show of pine cones. Courtney said it was not always easy to find these super trees. "Sometimes I could walk in the woods for two weeks without ever spotting one," he claimed, with his usual sparkle in his eye. "And it was also quite easy to get lost in those same woods," he added.

Courtney told those who ventured into the forest that they should remember to stop every so often and take a good look behind them as "that is what you will see when you are trying hard to return to your truck!"

Once when Courtney was looking for his super trees, he became lost in a huge Company timber tract in Gloucester County, Virginia, which was a very rare circumstance for him.

Claiborne Courtney takes students through his seed orchard, which was started in 1970.

He had tramped around in wide circles for so long he decided this time he was really lost. Finally he realized the way out of the woods was simply to follow the very trees he had earlier planted.

This plan worked for a short time until Courtney discovered the bad news—deer had nipped off some of his trees right to the core. The good news, however, was the deer had kindly left him a few seedlings every so often to offer clues to the way out.

Eddie Bernoski also remembered working in the woods to identify those super trees. "Once Dixon, Jones and I spent six months just walking the land and tagging all the super trees. I remember Courtney would shoot the twigs off the tree with a rifle. We used them as stock so we could graft them to seedlings already planted in a superior seed tree orchard. Always our goal was to find the fastest growing, tallest and straightest tree in the woods and then take the seed."

Courtney spent years developing super tree seed orchards with his devoted staff. "Pine grafting is almost like putting on a bandage," Courtney said with a laugh. "When it came to hiring just the right people to help me develop my seed orchards, I finally started to look for 'nurse types,' because caring for plants was very similar to caring for sick patients."

And the work could be just as intricate, too. Courtney finally realized that the best skills for grafting pine cuttings were the same kind of close work, finger-hand dexterity needed for seamstress or tailoring work. Some of his hardworking grafting crew were Eva Green, Mary Adams and Geneva Glazebrook.

Courtney liked to take students and anyone else he could latch on to through his pine

88

seed orchards and tell them all about the sex lives of pine trees. "Can you tell me how to tell a male pine from a female pine?" he asked me slyly during our interview.

I had to admit to Courtney I was stumped. Until I ran up on Chesapeake, I had come from the school that a tree was a tree was a tree. I had never realized any cross pollination and fertilization was going on out there in the woods among all those lively little pines.

It turned out according to Courtney's gleeful explanation that the male flower develops at the tip of the branch and the female flower develops about six inches from the tip. The interesting thing is pine trees develop both male and female flowers.

George Geron remembered working along with other technicians in those early seed orchards. "We fertilized, weeded, thinned, pruned roots and trimmed branches to keep those trees absolutely perfect," Geron said. Sharon Miller, retired chief of Woodlands, explained how the system worked. "Seed extracted from loblolly pine cones produced in these cross-pollinated orchards was used to grow superior pine seedlings in a Virginia Department of Forestry nursery in Providence Forge, Virginia. These one-year-old seedlings were then used to reforest Company lands with superior pine stock.

"Chesapeake orchards, like the orchard in New Kent County, Virginia, allowed Woodlands to show off its forestry skills to the rest of the world. There are no better looking pines anywhere in the nation than Chesapeake super trees," Geron said proudly.

Geron said it was a matter of pride to work to produce the Chesapeake super trees. Sometimes they could establish a super tree in the seed orchard just by grafting a super tree limb into a regular pine. They did this by splitting the tree down its middle, putting the limb inside and closing the wound to heal by using a simple rubber band. "It was really something that we could improve, a simple, everyday, regular pine tree," Geron said.

Like so many others who worked in improving pine stock, Courtney spent years in the

Celebrating during the 1980's are, from left, Sture Olsson, Bud Johnson, Tom Harris, Claiborne Courtney, Louise Courtney and Sharon Miller.

woods improving Chesapeake trees. Over the years he found many interesting objects. Displayed in his den in his West Point home is an impressive collection of Indian arrowheads and other primitive tools he found while working in the woods.

When Courtney retired, the Company thought so highly of his contributions that it dedicated the pine orchard in New Kent County to him. He still feels a great deal of pride when he drives by the "Claiborne Courtney Pine Seed Orchard."

"You know, only Elis Olsson, Sture Olsson and I had something named after us in the Company," Courtney said with a big smile. The new paper machine at the mill had been christened "the Viking" in honor of the old man and the two tugs had been named after Elis and Sture Olsson. Recently another treasured Company employee was so honored when the Tom Tyler nature trail in Eastern Shore opened in 1997.

There is no doubt about it. Courtney, with his very own tree orchard named after him, stands in very fine company.

Chapter 10

Land Acquisition

A paper company without its own private source of woodlands is like a king without a realm. It was fortunate that Chesapeake began a long-term land acquisition program as soon as it did. Because the purchasing began early, much of the woodlands now held by the Company were purchased at very attractive prices. Because of the proximity of these woodlands to the existing mill and to northern markets, they have become the solid foundation of great value in the Company.

In the early days the old man, Elis Olsson, upon first hearing about a tract of timberland that was for sale might well have hopped right into his black Cadillac and driven out to meet a farmer like a bird dog hot on the scent of fresh quail. If he liked what he saw and if he smelled profits, he would have paid the farmer cash right on the spot for the land.

Who knows? Maybe the two sons, Sture and Carl Olsson, on occasion made that same kind of land purchase. I heard plenty of tales during my interviews that some of the early land bought up by the Company was purchased during the depression for dirt cheap prices. But the closest I could come to any official knowledge of early land acquisition for the Company was my interview with Wyatt Mettauer "Tau" Crute of West Point, Virginia, retired director of land acquisition for Woodlands.

Crute was a New Kent County native who told me flat out he "had the good sense to marry a West Point girl." He firmly believed that nothing but trouble comes to a woodman when he marries someone outside of the "family."

It might also be said that there was no better name in the business than Tau Crute (unless it was Forrest Patton) for buying timber and land. Such a name had just the right sound to it, with the two quick, hard, one-syllable rings right off the bat, like the "bang, bang" of a gun aimed at the heart of some no-good critter. Who would fool with a man with a name like Tau Crute?

Crute's own grandfather had sold land to Chesapeake, so his memory takes us right back to Elis Olsson himself. Hired by Cecil Woodward in 1946, then manager of Woodlands, Crute started right off working for the Company by purchasing timberland with Andy Brooks, acquisition manager, who later retired in 1950.

"We must have purchased over 250,000 acres during my time with Chesapeake," Crute said. "Why, in those years I used to go over to the Eastern Shore and spend a week at a time just looking for available woodland." Crute remembered he stayed at the old Wicomico Hotel at Salisbury, Maryland, "where they had a good dining room," a very important ingredient for making a good land deal.

In those days before the Chesapeake Bay Bridge Tunnel, they used to travel over to the Eastern Shore by ferry from Little Creek in Norfolk or Ft. Monroe in Hampton, over to Cape Charles across the bay. The trip only took an hour and the big advantage was they could take their cars with them. Another means to the Shore was to take the ferry out of Annapolis, Maryland, but that took all day with the extra car travel required from West Point.

Eddie Bernoski remembered working with Tau Crute in the 1950s when he was out scouting for land. "That Crute would drive by a piece of land and know how much the wood was worth without even leaving his car!" Bernoski said.

Bernoski remembered he and his crew would then go over the piece of land and examine the timber stands in such detail that even such things as how close the tract was to the mill and what the transportation costs would be to move the wood were considered. All these details would finally go into the amount the Company figured the land was worth and then a bid was presented.

He and Crute, Bernoski recalled, would be looking at land and making a bid while other buyers from competing mills would still be walking over the tract! "We would park our car down the road from the tract and then 'cruise' from a safe distance without them even knowing we were there!"

During the 1960s, the Company was on a massive drive to buy land and there was often sharp competition from other interests for the same tracts. Bernoski remembered once being in Chase City, Virginia, hunting land buys with Vic Kaczmarski. The two wanted to keep it quiet that they were "Chesapeake men" in town looking for land to buy.

"It got out we were on some secret mission," Bernoski said with a laugh. "And like in all small towns, the news passed around fast that we were up to something. But we never told anyone who we were and this made their suspicions even worse."

Finally one day when Vic and Eddie were at the dry cleaners picking up some clothes, the manager came running out and said, "Listen here, you two guys, I want to know your names."

"Eddie looked at me and I looked at Eddie," Vic remembered. "Finally Eddie said, 'Oh, well, my name is George.'" But the manager, not happy with this, demanded more information. "George who?" he countered.

Eddie, not being much of a natural liar, hemmed and hawed trying to think of some plausible last name but his mind had gone blank. "So tell the man your last name," quipped Vic, enjoying the joke immensely.

" 'Er, George GEORGE,' " Eddie finally answered before bolting out the door, not a terribly original reply but the best he could do on such short notice. To this day, whenever Eddie and Vic see each other, they call each other "George."

Later on, the Company owned two airplanes that could take passengers over to the Eastern Shore and back. The new airplanes gave convenient service to the procurement office and travel was much easier for "cruising" distant sites.

Ironically, the growing need for air transportation throughout the Company may have played a part in sounding the death knell for the mill. Crute remembered one of the reasons Carter Fox, the CEO who sold the West Point mill in 1997, moved the Company Headquarters to Richmond back in 1988 was that management wanted to be nearer to the Richmond airport.

Crute came to know almost every piece of forest land in a 150-mile radius from the mill. He became a master at estimating timberland and had a knack for sometimes being able to evaluate the wood and land without even leaving his car!

He bought some pretty inexpensive land for the Company which sounds almost unbelievable by present day land valuation. "The first tract I bought for the Company was 100 acres over on Eastern Shore for $50 an acre," Crute remembered. Once he even picked up a tract of land in Buckingham County for just $20 an acre.

Some of that cheap land eventually went back to the community, however. Crute remembered the 300-plus acres of land Chesapeake gave to help build the new Colonial

Downs race track in New Kent County, land he had purchased for $40 an acre back in 1947.

George Geron, who had started work with the Company as a technician and ended up buying timber, remembered two ways the procurement office would purchase timber, depending on what best suited the owner. Either the company would buy the whole tract of timberland at an agreed lump sum or simply pay for the wood as the work crews cut it. The latter plan was good for the farmer because he knew exactly how much wood was going off his property and exactly how much money he was receiving for every load.

"In the old days when I would cruise timber for my old boss, Bill Rilee," Geron said, "I was authorized to spend up to $150,000 on my own for the purchase of timber cutting rights, which was a lot of money in those days. We also worked to sell our Chesapeake Forestry Action Program (CFA), which was a plan Chesapeake developed with private landowners where we would have first refusal on any timber they would sell in exchange for the Company providing them free seedlings and management services."

This CFA plan became very popular with farmers across the state and even Carter Fox, while CEO of the Company, held land in the CFA program. "Within just ten years of starting CFA, we had over 50,000 acres signed up," Geron said.

Geron remembered that over the years he occasionally had troubles with the landowners. "One year we bought the timber off a private tract in Richmond County. When we showed up to cut the timber, the wife came running out of the house terribly upset to hear we were actually going to cut, burn and reforest her timberland." Geron laughed at the memory of the woman falling to her knees, crying and begging them not to cut her darling trees. "We finally called in the state forester to explain to the lady the trees were mature now and needed to be harvested or they would die on their own and be useless to everybody and she surely would not want that to happen. So the best thing to do, we told her, was cut them and plant new trees for the future," Geron said. "She still cried. We ended up having to cancel our deal with her and then she went and sold the land to somebody else!

"You could get into all sorts of troubles with relatives and land boundaries, too," Geron said. Once he made a deal with a woman in Mathews County, Virginia, to harvest a piece of her land. Her father, who lived next door, came out to ask questions. "It turned out he was part owner of some of the land. She got so mad that I had talked to her father, she called the Company to complain and even wanted me fired!" Geron laughed.

But Geron found a good way to test property lines whenever there was any question of ownership. "I just cranked up my power saw and started to work along the boundaries and, sure enough, the real landowners came running out real fast to tell me exactly what they owned!"

Crute retired in the mid 1970s. By then, chief forester Dick Brake, who had originally come to Chesapeake in 1960 one week out of forestry college, took over the responsibility of land and timber acquisition.

Brake, whom Tom Harris had always called "one of my stars," came to the Company at a time when there were only four foresters in Woodlands. Brake remembered first coming to West Point as a single man right out of school in a 1953 Chevy packed with everything he owned. He stayed in a room down on First Street in West Point. When he got the assignment to Lynchburg, Brake decided this was the time to get married.

"I borrowed $800 from the Company credit union and Jimmy Sears co-signed for the loan," Brake said with a laugh. "When Carl Olsson heard I was getting married, he came out to the parking lot and looked at my old, beat up jalopy and shook his head. 'You're going to take a Company car for your honeymoon,' he told me, 'because I want you to come back in one piece!'"

Brake remembered Carl Olsson as an awfully nice fellow, smaller than his brother Sture. A funny memory Brake had about Carl was every now and then he would lean back so far in his office chair, it would topple over backwards with him still in it. Later, Tom Harris, who took over Woodlands, had the same trouble with this chair.

Finally Harris had enough of falling backwards, sent the chair to the warehouse for storage and arranged to get a new one, one that would have enough sense to stay in place with a big shot executive sitting in it. That toppling chair came back to Harris, however. When he retired, his staff had it removed from storage and wrapped up as a nice farewell gift to their boss.

Carl Olsson had announced in the early 1960s the Company wanted to save its wood close to the mill for possible future use. Brake was sent to the Lynchburg, Virginia, area to arrange to bring in wood from the mid-state area. Brake worked closely with Bob Sales and other wood dealers in the area in the task of bringing in the wood.

By 1964, Brake was named chief forester for Chesapeake. He believed he got that position because he had drawn up a plan of recommendations concerning the Company's land holdings with ideas on a reforestation program.

In 1970, Brake wrote a timely report identifying Company land that had special value for the Chesapeake Board of Directors. With his knowledge of the special value of Company land, it was natural that Brake was appointed in charge of land acquisition.

As Brake continued to acquire land for the Company, one thing was very clear in his mind. He knew Chesapeake Corporation had some exceedingly valuable land in its fold, some of it beautiful waterfront property. This land was too valuable to continue to hold just for wood production. Brake was the only person in those years in the Company who had personally walked over every piece of Company land. He knew the holdings well, what the waterfront views were, even the depth of the water, and also their true market value.

One of Brake's land acquisition trips for the Company took him on a reconnaissance of St. Joe Paper Company's lands in the Panama City-West Bay area of Florida. The year was 1970 and the month Brake and Forrest Patton went down to Florida just happened to be February, probably while the snow and sleet were knee deep

The famous trip to Florida resulted in this car carrying foresters Brake and Patton becoming stuck the middle of St. Joe Paper Company's million acres of timberland. Brake, in the distance down the road, hiked out to get help while Patton laughed and took pictures.

back home in West Point. One trait of a good land cruiser is to know the right time of year to be out cruising land in Florida!

An occurrence that was destined to become a "classic Chesapeake tale" took place at this time. While the two foresters were cruising Florida, Brake kept noticing orchards of a certain kind of nut trees. "I wonder what those nuts are," Brake said to Patton. Finally, all those nuts were just too tempting and Brake pulled up next to an orchard. He got out of the car, picked a nut and looked it over but could not identify it. Patton also examined the nut and had to admit he had never seen a nut like it.

Curious, Brake bit into it. "Tastes pretty good," Brake said. Patton shook his head and advised Brake he'd better not risk eating any more of it. But Brake, being the natural risk taker he is, finished the nut and the two men went merrily on their way.

Soon they saw a country store and stopped off for some Cokes. "Say," Brake asked the clerk at the counter, "what kind of nuts are growing in all those orchards out there?"

"Tung nuts," the man said. "They're used for paint. And you better not eat any either or you'll get real sick."

Brake admitted he had tried a tung nut and found it to be quite delicious. "Better drink a quart of milk right away," the man advised grimly.

Later the foresters stopped in a motel for the night and went to a restaurant for a dinner of grilled steaks. Everything went fine until the two platters of steaks arrived steaming hot to the table. The tung nut must have reached its goal because Brake suddenly felt queasy and ran for his room. Brake was so sick from that one nut that he spent an hour on the commode while at the same time throwing up in the tub.

An hour later a very weak Brake returned to his table. "Where's my steak?" he asked Patton, while looking down at his empty platter. "Ate it," came the swift reply. All these years later, Woodlands Division howls laughing when anyone mentions a tung nut.

"We looked over a million acres of Florida timberlands," Brake remembered of that trip, probably the biggest land package ever "cruised" by Chesapeake men. Brake already had an idea that the Florida land package had tremendous potential for real estate development, yet the Company never made a bid on it. Nor did the Florida land ever sell in all those years until recently when a real estate developer bought some of it.

By this time change was in the wind and Brake knew it. He was beginning to question how much longer Chesapeake would still be interested in making kraft paper. He foresaw the Company moving into other areas of profit and using their huge land holdings in far more profitable ways.

"By 1981 I updated my original 'special interest land' report to the Board of Directors," Brake said. "In the meantime, I was busy taking courses in real estate and law at night school at the University of Richmond." It was as though Brake were preparing for the changes that lay ahead.

By 1983, Brake was made the first president of Delmarva Properties, which was Chesapeake's first land development subsidiary. Under Brake's leadership, 2000 acres of land in James City County had grown to a tract of now over 7000 acres. This is developing into a beautiful waterfront community called "Stonehouse." This development is only one of many residential areas on land purchased as Chesapeake timberlands.

Another big Chesapeake land deal was the actual outright gift of 365 acres to Colonial Downs Race Track in New Kent County. The Company deemed it most important to encour-

From left are Robert Geron, Dick Brake, Walter Zingelmann and Mary Ann Fetterolf.

age the racetrack because such an enterprise would draw other businesses to the area, creating new jobs and tax revenues. Because Chesapeake owned the 7000 acres surrounding the racetrack tract, this gift was certainly considered an excellent long-term investment for the Company.

According to Jim Vadas, the Company has sold 25,000 acres of non-strategic timberlands but continues to hold about 329,000 acres: 80,000 on the Shore, 104,000 in the Piedmont area, and 145,000 in the West Point area. When the paper mill at West Point was sold to St. Laurent Paperboard Corporation in 1997, Chesapeake held on to all its timberland. (This land was sold to Hancock Timber Resources Group in 1999.)

The early procurement of large land holdings generated another source of revenue that perhaps early purchasers had never envisaged—hunting fees. Chesapeake now collects over one million dollars each year from hunters who lease Chesapeake land for sport. The timberlands also gave Chesapeake an opportunity to provide summer intern positions for many forestry students who need actual land management experience in on-the-job training.

Timberland requires constant maintenance to keep it producing healthy, good quality wood, and the interns learned and earned while Chesapeake reaped the benefits. As an added bonus, Chesapeake used this intern program as a great way to assess which students they wanted to hire as foresters one day.

Chapter 11

The Eastern Shore

Soon after this book project was started, Chesapeake's operations manager, Jim Vadas, called to tell me I was scheduled on the next morning's flight to Salisbury, Maryland. While he and a staff member from the Company's Human Resource Department did some group interviewing on hiring a prospective new forester for the region, I was booked to interview some of Chesapeake's Eastern Shore troops.

Vadas thus delivered a very important point to his newly hired author. I had interviewed past Woodlands chiefs Tom Harris and Sharon Miller just one week before. Number three interview on the list would start on Eastern Shore. I realized from that point on the importance the Company placed on Eastern Shore, all its employees working over there and its large land holdings.

It was a beautiful, sunny, blue-sky kind of day as we took off from West Point Airport in an eight-seater turbo-prop with two pilots at the controls. The drive from my home in Urbanna to the regional airport in Mattaponi, Virginia, right outside West Point, had been just 20 minutes and there was no such thing as a parking problem. I just pulled up next to the airplane hangar and boarded the plane, which was waiting no more than a few yards away.

The trip lasted less than a half hour. It was spent with me staring out the window in rapt fascination at the broad Chesapeake Bay spread beneath me like a big blue carpet. Tiny tugs pushing or pulling barges and an occasional freighter pushing out a stream of wake provided interesting dots and dashes of white across the blue sea. Did I see the "Sture" or the "Elis" chugging across the bay and piloted by Captain Bo Traywick of Deltaville, pulling two barges of chips with her nose hell-bent for the mill? I imagined so.

Chesapeake's "classic" forester, Tom Tyler, was waiting for me in Salisbury airport with his pickup truck. Already sick from cancer, sick enough not to be able to eat the lunch he ordered with me later in the day, this handsome hulk of a man took me under his wing and told me all he knew about the business of bringing in the wood.

"You're not one of those Episcopalian lady society types, are you?" he asked with his charming crooked smile. I threw back my head and laughed. I liked Tom Tyler, really liked him, from the moment I set eyes on him because he spoke his mind. I'm a writer and I value, almost more than food the real, honest, salt-of-the-earth kind of people who speak what they think without a lot of trappings.

After assuring him I was no society lady type of any church background, he seemed cheered considerably. We zoomed off in his truck to the local cafe to meet with Bob Owens, retired manager of the Shore; J. V. Wells of the J. V. Wells Lumber Company, who provided the Company with at least five truckloads of chips each week; and Larry Walton, present manager on the Eastern Shore. Sam Owens, son of Avery Owens, a pulpwood dealer who had recently passed away, was good enough to meet with me, also. With this representation, I quickly learned the special challenges and rich rewards that Eastern Shore provided the Company.

Bobby Owens, who had started work with the Company in 1951 and retired in 1989, had seen almost 40 years of service. In all those years, there was nothing he had not run across concerning problems with bringing in the wood.

Owens perfectly fit the mold for a Chesapeake man. After service in World War II, he began work for his dad on the Shore in the grocery business but loved working outside and was soon buying up timberland to resell at a profit to the sawmills. Eventually he ran into Gerald Harris whom Owens remembered as the first Chesapeake wood dealer in Maryland. Harris told him Red Highland back in West Point was looking for someone to represent Chesapeake's interests on the Shore.

That was all it took to get Owens on board. Soon he was on the Chesapeake payroll and his first big assignment was to buy land for the Company. "I bought over 90,000 acres at the time when no one else was buying timberland," Owens said, "and plenty of that land I could get for $60 to $100 an acre." Today Owens thought that same land had grown in value to almost $1000 or more per acre, a figure many would feel helped make Owens' lifetime work for the Company worth his weight in pure diamond-and-ruby-studded gold.

The next step was to begin harvesting the wood off that land. Back in those days there were not many environmental laws regarding wetlands, but Chesapeake people were already aware of some of the problems that harvesting pine off low land would generate.

"We tried to protect the land as best we could along with using it to our best advantage," Owens said. "For example, we already knew to leave a certain number of trees on a tract for nesting, wildlife and soil protection." This good-sense practice later became a requirement by federal and state law.

Owens was promoted to procurement supervisor in 1955 and his work mostly encompassed working with wood dealers in the Sharptown, Maryland, area preparing barge loads of pulpwood that had been cut in five-foot lengths to be shipped to the mill. The arrangement worked so well for the Company that they bought needed quotas of wood from private sources and conserved their own wood supplies for later use. This policy lasted until 1968.

In 1968, as a part of Tom Harris's new area districting plan, Owens became manager for Area 2, which was the Eastern Shore of Maryland, Delaware and Virginia. His duties were pulpwood and log procurement, land acquisition, and land management within this area. It was quite a job assignment. Owens suddenly found he had new responsibility up to his ears.

There is a favorite story in Woodlands concerning Bob Owens. Once while inspecting a tract of timber Tom Tyler ran into what Owens called "a real mean farmer." Tyler knew Chesapeake had cut some of this farmer's timber by mistake and that this farmer had a tough reputation. On top of that, Tyler noted the farmer wore a holster packed with a pistol and he figured the gun was definitely loaded.

Finally Tyler managed to calm the farmer down promising to make full restitution for any damages. The farmer said, "OK, I'll accept a cash settlement, but if I ever meet that owner of Chesapeake Corporation, I'm goin' to kill that son of a bitch right on the spot!"

With this remark fresh in his mind, Tyler took Owens over to meet that same farmer. Tyler turned to the farmer and as smooth as a hot knife slicing butter and pointed to Owens, "I want you to meet Mr. Owens, the owner of Chesapeake!" Even the "real mean farmer" had to laugh.

On top of Owens' regular responsibilities as area manager, Chesapeake Bay Plywood Company, a joint venture between Champion International and Chesapeake, had started up in 1965 over in Pocomoke, Maryland. It added additional stress to an already high stress level for Shore personnel.

"We had total responsibility to provide 30 million board feet of plywood 'peeler' logs a year to this new company," Owens said, "along with our weekly quotas of 600,000 board feet of debarked peeler logs and 2,000 cords each week of chipped residue over to the paper mill in West Point."

Owens said some of the big wood dealers who worked with him to meet those goals were H. V. "Danny" McKay, Avery Owens, Lassiter Lumber Co., Dave Belote and Charlie Knoller. Their efforts were greatly appreciated.

The new Chesapeake Bay Plywood Company produced its very first sheet of plywood on May 27, 1966. Owens remembered the event as a big one for the Company. But with the added, growing need for wood at this new facility, "We were really under the gun to provide more and more wood," Owens said.

Other concerns were the new environmental laws that went into effect in 1972 which spelled out specific new rules and regulations for the use of wetlands. "For years we had troubles with the Environmental Protection Agency, EPA, which was questioning us constantly about our barges in the canals and any possible runoffs to the water, any chips falling in the water, any illegal dumping and so forth," Owens remembered. "But with Tom Tyler's help, we finally reached an understanding with them and dedicated ourselves to doing the work exactly right in order to protect the land and water. We took no short cuts."

After almost 20 years, Chesapeake Bay Plywood Company closed down because of economic problems that were worsened by a work stoppage called by the Carpenter's and Woodworking Union. This was the last straw for Chesapeake, which had dissolved the partnership with Champion and owned the plywood plant outright. The Company still runs its chip mill at this site but the plywood plant was eventually closed.

The constant stress finally took its toll. In 1979, Owens suffered a heart attack. His doctor insisted that he leave work every day and rest a few hours.

Owens remembered meeting Carter Fox about this time, who was visiting Pocomoke from the Richmond Headquarters. Owens apparently did not realize Fox was at the office and continued his business without any particular fanfare. Fox stopped to see Owens busy at his desk.

"Are you still taking a rest every afternoon?" Fox asked him. Owens looked up from his papers at the young man, hard like a weasel might look at a mouse, and answered, "Yes, son, and maybe if you stay as long as I have with Chesapeake and work as hard as I do, you can take a rest every day after lunch, too."

And that was the end of the matter.

Tom Tyler was born on May 27, 1941. He joined the Company after serving in the Navy for 4 years and in various other odd jobs here and there. "I ran into Chesapeake Personnel Manager Graham Evans and I told him I hated wearing suits and ties and I really wanted to work in the woods," Tyler said with a measure of passion, just the way he said everything.

Tom Tyler of the Eastern Shore, the "heart and soul" of Woodlands.

Tyler started to work for the Company in 1966 as a "forestry field man" and he went right to work in land management duties, planting trees, installing culverts, and clearing land. Tyler had a certain winning way about him. To know him was to like him and he was soon supervising work crews and contracting with loggers for harvesting Company wood. Before long he was supervising 18 logging crews. That may have been too much stress because he suddenly up and left Chesapeake and went to work for the sheriff's department in Dorchester, Maryland.

What a change from the woods to the city. As a cop, Tyler saw the worst troubles of society including the H. Rap Brown race riots that left a lifelong unfavorable impression on him. The rising number of race riots, along with one year straight on night shift, soon sent the policeman back to Chesapeake. Tyler decided supervising 18 logging crews for Chesapeake was not so bad after all.

Working as Bob Owens' right hand man, there was not much that Tyler was not doing on the Shore. His duties varied with the day of the week. Tyler was a forester for all seasons who could do anything, from collecting hunting fees from various hunting clubs who leased Chesapeake lands to writing land contracts for logging companies to harvesting timber and to planting and managing trees.

Before long Tyler was a fully licensed forester in the state of Maryland. He was very proud of the fact that he and Bob Owens were the only licensed foresters in Maryland who did not have a university forestry degree. But Tyler was quick to add this does not happen anymore.

Tyler's main duty was the overall care and maintenance of over 17,000 acres of Company land. As Tyler worked on the land, something interesting happened. He described the phenomenon in a simple statement, "The more I cared for the land, the more I cared for the land." Thus, Tyler quite naturally became a conservationist, making sure his land was managed correctly, always having buffers along the shore and streams, caring for original homesites and cemeteries found on "his" land, and taking care of both the land and wildlife in a conscientious manner.

Tyler loved wildlife and was personally involved with the care of much of it. A favorite memory from Woodlands chief Paul Harper concerned Tyler's hobby of raising little wild ducklings the state of Maryland provided him. Tyler cared for the ducks to adulthood and finally set them free in his pond right in front of his house.

Harper, who loved "duck huntin'" like most of the Chesapeake crowd, remembered that every time he was over on the Shore, he would stop off at Tyler's house, see the wild ducks and rub his hands in glee. "Can't wait to shoot those ducks!" Harper would say to Tyler just to watch his reaction.

There would be a long pause in the conversation. Tyler would then respond in a steely tone, "Don't you dare shoot my Huey, Dewie, and Louie!"

Once when Harper was with Tyler and his ducks, he noticed Tyler had a big switch and was walking after them "encouraging" them to leave the "reservation," take flight and seek life in the wilds. One big old duck, used to the daily handout of free corn, followed Tyler around all day long absolutely refusing to budge, regardless of that testy little switch.

"Tom chased that duck all over the place," Harper remembered. "Finally the duck got the message and took flight. He immediately hit Tom's clothesline and landed kerplunk right under it. That was that. The duck was no fool. He knew immediately that ducks

Tom Tyler, Elois Brown and an undentified worker on the Eastern Shore.

101

weren't meant to fly. No matter what Tom did with his testy little switch, that duck never flew again!"

As environmental interests rose in the general public, Tyler decided to get actively involved at the grass roots level. Before long he was attending environmental meetings and associating himself with various "green groups." Tyler was never afraid to stand up and state his point even under scrutiny of the most radical groups. Slowly and surely Tyler won the respect and trust of many of these groups. Tyler provided the foundation of a long and close relationship between Chesapeake, the EPA and various other preservation groups.

Tyler also got into politics and made sure the voice of real woods experience and good common sense was heard on all environmental issues. Tom was a valued member of many environmental groups in Maryland including the Government Environmental Advisory Committee, Dorchester County Forestry Board, Maryland Park Foundation, past president of the Forestry Resource Conservation and Development Board, the Chesapeake Bay Foundation and even served as president of the Maryland Forestry Association.

Tyler was a one-of-a-kind person, the sort of employee companies really appreciate because he was honest, loyal, hardworking, spoke his mind, and stood up for what was right, no matter what. This is a breed of man that is slowly disappearing from the corporate scene. Tyler often spoke out on controversial and sensitive environmental issues and his opinions were sometimes picked up in the Baltimore Sun and other newspapers.

Some of the staff at Chesapeake were a bit concerned about Tyler's unusual relationship with the environmental groups, government and media. Once Tyler was particularly outspoken to a reporter who later published an article giving Tyler headline coverage in the Baltimore Sun.

Almost immediately some flak came back to Tyler from Company Headquarters. Maybe Tyler should not be speaking to the press about such sensitive matters, some advised. Then CEO Carter Fox saw the article at Corporate Headquarters in Richmond.

Tyler remembered the call that came right in to his Sharptown office on the Nanticoke River as he was sitting at his desk sipping a mug of coffee. "That article you helped write was the best damn story I have ever read on the subject!" Fox shouted enthusiastically over the telephone. From that point on Tyler never once hesitated to speak out on important issues to the press.

Tyler was a tremendous practical joker. He loved to play tricks on people and one person he especially loved to snag was nobody other than the Company surveyor, Dick Cartwright.

One time Cartwright was riding around the shore with Tyler when they happened to spot a cornfield that Tyler knew was owned by a particularly ornery farmer who had a reputation for watching his fields of corn with an eagle eye. He was even known to come out with a blazing shotgun anytime he saw someone taking any of his corn.

Tyler slyly pulled up to the cornfield in full view of the farmhouse so as to be sure the farmer could see him. Then he turned to Cartwright and told him he knew this particular farmer and he really liked Chesapeake people. "Anytime any of us ever want any corn, we can just help ourselves," Tyler said with a crooked smile.

Cartwright took one look at the beautiful rows of green corn and joyfully hopped out of the car. When he came out of the field with his arms filled with as many ears as he could manage, he was surprised to see what certainly appeared to be a most irate farmer running toward him from the farmhouse. Cartwright also saw the gun. He dropped the corn and ran for the car.

But before getting away Cartwright heard the farmer shouting at the top of his lungs. "If I ever see you in my cornfield again, I'll have you arrested!" And then he uttered two more words on the stern of Cartwright's getaway car just to illustrate his point, "BANG BANG!"

Another time Cartwright and Tyler were up in Maryland and they happened to drive through the town where Tyler was once deputy sheriff. Tyler stopped and made a quick phone call to a buddy still working on the force. When he got back to his car, a local deputy had already pulled up and was talking sternly to Cartwright.

"Are you Dick Cartwright of the Chesapeake Company?" asked the deputy. Dick swallowed, then said respectfully, "Why, er . . . yes, sir, I am." (He must have wondered if the IRS was still after him.)

"Well, then, I have a warrant for your arrest," the deputy said snapping on handcuffs. Cartwright was incredulous as he was led off to jail. "Don't I get to make a phone call?" he asked as he was booked.

Then Cartwright heard someone laughing hysterically behind the door. He turned around and there was a jubilant Tom Tyler. It was almost as if Tyler had had a longtime dream of getting Cartwright thrown into jail and, at last, his dream was about to come true.

By the time I met Tyler, he was suffering from lung cancer that had spread to his shoulder. He was fighting for his life. The day I spent with him, he was already struggling with the symptom that all cancer victims eventually experience through both the disease and the treatment, the slow, insidious process of losing weight. His once big and muscular body was a thing of the past, leaving a thin frame which did not quite suit the rugged forester that everybody knew and loved. Along with the weight loss, the chemotherapy was taking its toll on his general feeling of well-being.

"I'm going to lick this cancer," he told me. I looked him in the eye, like writers do as if to read a man's soul, and saw the fighting spirit that Tom Tyler possessed deep in his heart and soul. Then I willed back the tears that started to well up in my eyes. How I wanted to believe him. How I hoped he would win. How I liked this passionate Chesapeake man.

Tyler died less than a year later. A contingency from West Point flew over to Eastern Shore to pay him their last respects. He was well loved, so much so that when Woodlands chief Jack King talked about Tom Tyler in an interview several months later, King broke down and cried. "I loved that man," King said.

Tyler had a rare ability to bond with top people in government, which had made him especially helpful to the Company. "Like Bud Johnson had close connections with the top political contacts in Virginia, Tom Tyler had those same contacts in Maryland," King told me. "If you needed the Maryland governor for any reason, we all knew all we had to do was to call on Tom."

Out of absolute love and devotion to Tyler, the Company gave him its highest honor. It finished a nature trail Tyler had started talking about over 30 years ago and had actually started developing at Mardella Springs along the Nanticoke River. Angela Hall, forester on the Shore, told me Chesapeake employees volunteered their time and energy to finish this trail in memory of Tom Tyler. "It was easy to do," Hall said. "We just followed the flags Tom had already marked out for us in the woods."

The trail weaves through some of the most beautiful land on the Eastern Shore in a

A tribute to a citizen of the bay

■ **Forester:** *Always willing to meet environmentalists halfway, Tom Tyler leaves a legacy of passion for the Shore's rivers and woods.*

By TOM HORTON
SUN STAFF

SUN STAFF

BIKING NEAR HOME the other afternoon, I passed a forest. Loggers were at work, taking it down. I waved a greeting. They are doing a pretty decent job, I thought.

In an ideal world, I like my trees all standing, my forests undisturbed. The sound of a chain saw can spoil my day.

I said as much to Tom Tyler shortly after I met him a few years ago. "Where the hell you think that paper you're scribbling on comes from?" he shot back.

The sparring was good-natured, but with a dead-serious undercurrent. Tom was working overtime to determine how dangerous we environmentalists were. We were trying to figure out whether he was for real.

"We" were an assortment of academics, river rats and local, state and national environmental groups who had coalesced around preserving the rural character and ecological diversity of the Nanticoke River and its watershed.

Tom was an umpteenth-generation Dorchester countian who'd fished and hunted the Nanticoke region for most of his 55 years.

He was also regional forester for Chesapeake Forest Products, whose timber holdings up and down the Nanticoke made it the region's largest private land owner, a key to the river's future.

Even more critical, Tom had the respect and the ear of farmers, politicians and other rural residents the length of the river.

It wouldn't have taken him more than a few phone calls to foment an instant property rights uprising against the likes of the Chesapeake Bay Foundation, the National Park Service and the rest of us, delaying or killing any Nanticoke environmental programs.

In the end, Tom Tyler did more to start a truly effective effort for the river than anyone else could have—so much so that I suppose we'll have to name some place along it in his honor one day.

I'd rather not be thinking in terms of memorials: would rather have Tom still with us smoking too much, eating too many sausage sandwiches in every diner and country store between Accomack and Sharptown; would rather still be putting up with the calls he delighted in making to us "greenies" at daybreak on the cell phone in his Ford pickup as he sped among forestry projects.

But a cancer, discovered last winter, took him the evening of Aug. 12. He was only 55.

He went way too soon—not only for his family and friends, but for the rivers and woods of the lower Shore.

Tom was among a handful of extraordinary citizens of the bay I've come to know in two decades of environmental writing. He had a fierce passion for his region and for the traditional work of its people; also, he grasped that a larger view could help maintain the region's environment in changing times.

After Nanticoke environmental meetings, he would sit in his pickup and pepper me with questions about the bay foundation, the Nature Conservancy, various individuals.

Where was so and so coming from? Did this person represent what his group felt? Did I trust that person? And I'm quite sure he was asking as much of others about me.

I also knew that his willingness to meet environmentals halfway was not without risk. Some lifelong friends in rural Dorchester had turned on him for his efforts.

One night, Tom suggested that I meet him at his office the next morning in Sharptown; we'd take a walk in the woods.

On his office wall was a poem about the Nanticoke, written by an 11-year-old who'd come through on a bay foundation canoe trip. Tom was letting the group use an old lumber yard there as a campsite.

Pinned next to that was an article Tom liked, "A Deal From Hell," by some property-rights guy out of northern Maine, about how conservation easements were part of environmentalists' hidden agenda to take land out of production and erode freedom.

What did I think of the article? Tom asked. I said I thought it was easy to write from the Maine woods, that if the writer were in fast-suburbanizing Maryland he'd sing a different tune.

Although Tom bitterly opposed any notions of making the Nanticoke a national park or wilderness preserve, I had noticed that he was at least as upset as any environmentalist about the sprawl development that was encroaching on the Shore's farms and forestland.

"A goddamned cancer, and wherever our [Chesapeake Forest Products] property stops is where it starts," he would say.

And I had to admit, in the real world, commercial tree-cutting and replanting often seemed the best alternative.

That day Tom showed me his pride in Chesapeake's cutting practices, which left larger-than-required buffers of trees standing along roadways for scenic purposes, and along waterways to preserve water quality and wildlife habitat.

He was equally proud of "all the places we do not cut," including some of the state's rarest old Atlantic white cedars, where he was laying out a nature trail along the Nanticoke.

"Don't the woods just smell good?" he asked as we walked through myrtle and among pines, laurel, magnolias and cedars.

We can keep the river healthy without making it some kind of national park and shooing out the farmers and loggers and watermen, Tom said.

Let's just shoot for keeping it so that if people want a national park a hundred years from now, we'll still have the option, I said.

At his funeral, hundreds of people turned out—the forestry community, politicians conservative and liberal, rural and urban; environmentalists; and natural resources officials.

He cast a wide net. Always, people said, Tom would tell them exactly what he thought, and he would always ask them, straight out, what they thought.

And, above what anyone thought, Tom Tyler cared for the rivers and woods.

This article appeared in the Baltimore Sun after Tom Tyler's death.

meandering two-mile boardwalk. It even sports an information center with brochures for self-guided tours. All the different trees and shrubbery are identified so the public can learn their names and characteristics. The range includes sweet gum, ground (Princess) pine or running cedar, southern red oak, American holly, white oak, black gum, Virginia pine, loblolly pine, pond pine, white cedar, sassafras, goat's rue, mountain laurel, shad bush, high bush blueberry, and yellow poplar.

In addition to the trees and plants on this 455 acre site, the public can view a real swamp, turtle egg sites, an old homesite and cemetery of an early Eastern Shore farming family, and a variety of teeming wildlife.

Tom Tyler would have been proud.

Chapter 12

The North Carolina Connection

Although this book revolves around many people who lived and worked in West Point, Virginia, it is important to realize that most of the work in procurement of wood for Chesapeake was taking place very far away from the paper mill. Company business took place over a three-or four-state area. Jack King, chief of Woodlands today, estimated that over the years at least "two thirds of everything we did for the Company happened away from the West Point mill.

"Sometimes the West Point people did not seem to realize just how spread out we were," King added with a chuckle. "I used to remember talking to my boss back at Woodlands when I was working in Keysville, Virginia, in the middle of the state, and they would say to me . . . 'Well, I guess we can't do such and such today because it's raining too hard,' and I would answer . . . 'But it isn't raining over here in Keysville!'"

The Company produced its number of very unique people. Or, perhaps it did not produce them so much as it had a knack for hiring them. Elis Olsson, who founded the Company, and his sons, Sture and Carl, had to be on the top of the list of special people. It follows that such people would attract other special people. If I were compiling a list of special people within the Company, Bob Sales, Claiborne Courtney, Jimmy Sears, Tom Tyler, Chief Webster "Little Eagle" Custalow, Randolph Fogg and Dick Cartwright would certainly be on that list.

And no list would be complete without mention of Bill Ellison, better known as the "Bull," which Jack King explained was a nickname ascribed to him in honor of North Carolina State University and Coach Edwards, better known as "Three yards and a cloud of dust." For a nickname, the "Bull" fit just right. The name recorded his style.

Although Ellison died of a heart attack in 1983, Jack King, Jim Vadas and Larry Walton well remembered the Bull, especially for this book. (The Bull was famous for suffering 25 broken bones during his lifetime, a fact that he often bragged about to his new foresters.) Perhaps these Company new hires were able to describe life with the Bull so well because Larry Walton, Jim Vadas, Steve Whaley, Wayne Beasley and others, besides earning a regular forestry degree from an accredited university, were all graduates of the "Bull's School of Wood Procurement."

King, who later became vice president of Woodlands, remembered life in Area 6 in the Henderson, North Carolina, district, which was once "governed" by Ellison. King was then area manager in the Keysville area immediately adjacent to the Bull's operating arena. The two men developed a special kinship in commiserating over their similar hardships and problems.

A graduate in forestry from North Carolina State, King had worked for the North Carolina Forest Service before coming on board with Chesapeake in 1966. His first assignment was working under Bob Owens at Eastern Shore (which some foresters would say was in the same league as the Bull's training program).

King remembered the horrible first winter he came to work for Chesapeake and being sent over to the bleak Eastern Shore, not exactly a Nag's Head resort. "It snowed so bad that year the ice never melted once from the windshield of my car but Bobby Owens never let bad weather get in the way of bringing in the wood," King remembered. Thus King had been well indoctrinated into Chesapeake's work style.

When King got to Keysville he began to hear about the Bull's infamous "Bill of Rights" for all new employees. "The first rule in the Bull's Bill of Rights was a new employee had to work 24 hours a day if he had to, in order to fill the wood racks," King said. "The second rule was 'If anyone isn't attuned to this kind of work ethic, it would be better if he left NOW.' " As one might ascertain from such rules, there was not a whole lot of give and take in the Bull's new employee's handbook.

One day while visiting the Bull in the North Carolina office, King, who had just taken over as manager of Keysville, decided to ask Ellison about the holiday schedule. He wondered if it would be similar to Bob Owens's holiday schedule, which consisted of a few days off for Christmas if they weren't too busy. "So what kind of vacation time can I expect over here?" King asked Ellison.

"I don't know," the Bull answered abruptly. "I never had one."

"The Bull," Bill Ellison, enjoying some fried chicken at a Company gathering.

That is the sort of response that can discourage a new forester to the "western" areas. But not King. He made sure the Bull's answer became a classic Company joke, especially for the new foresters just coming on board. The idea was that it was better to laugh than cry.

Until tragedy struck. The Bull died of a heart attack in a motel room away from home on Company business at just 57 years old. His premature death got everyone's attention in the Company. "Now we all know to plan for rest and a little regular getting away from the daily stress of our jobs," King said.

Another forester, Ed Tokarz from West Point, noticed in 1971 a big change as soon as he passed over the Virginia state line into "Tar Heel" country. That was the year he was sent down to North Carolina on assignment.

"In Virginia, we foresters always wore a coat and tie," Tokarz explained. But one look at Torkarz' gentlemanly attire and the Bull scoffed. "We don't wear any coat and ties to work in North Carolina," he grunted.

There was nothing for the Virginian to do but take off his coat and tie. Torkarz must have decided that when in Rome, a forester does like the Romans.

Forester Larry Walton, now area manager at Eastern Shore, remembered starting work and reporting to the Bull in 1974. "Since Ellison was home recovering from an initial heart attack, Jim Vadas, who was acting manager there at the time, became my immediate boss. Consequently, although my indoctrination to the famous 'Bull's Bill of Rights' was postponed a bit, I did not manage to dodge the eventual initiation.

"On Ellison's very first day back to work, he asked me why I wanted to be a procurement forester," Walton said with a laugh. "I had no real reason other than my professors at Clemson suggested there were more job opportunities for procurement foresters. So that became my paltry answer.

"Ellison stared at me and then said, 'Hummmmph! Wrong answer!'" Walton said. "Then he held out his hand high off his desk and said it would take a book this tall to explain to me what a procurement forester is all about."

Walton said Ellison then recounted the first two "Bull's Bill of Rights" but elaborated a little. "This job," he told me, "takes 24 hours a day, seven days a week, 52 weeks a year, or 8760 hours a year, and 24 extra hours every leap year!"

Walton remembered hearing more tenets in the famous "Bill of Rights," such as "Vacations are a privilege and not a right," and "Ask not what your Company can do for you, but what you can do for your Company." (The latter sounded a bit like JFK's famous remark. The president no doubt borrowed it from the Bull.)

"We soon learned other red flag subjects," Walton said, "like never, never ask when or if you would be getting a salary increase or be prepared for quite a lengthy soliloquy. And never, never say you thought you had proved yourself over time because that kind of remark would trigger a certain . . . 'You are never done proving yourself around here!'"

Jim Vadas heard many variations of those same responses. One of his favorites from the Bull was his classic remark, "Always remember, you are always being watched and tested."

One time Walton tried to argue the point with the Bull by reminding him that he had been working for him for four years. "Ha!" the Bull answered with a laugh, "What is a mere four years? Just 208 weeks, 8760 hours in a year, seven days in a week, 24 hours a day, and 24 extra hours every leap year."

Walton said the subject of vacations had to be approached so delicately that he never remembered actually taking a vacation in all six years he spent with the Bull. And only a fool would be so dimwitted as to ever tell Ellison about any personal interests or extracurricular activities like Jaycees or softball. The Bull thought any such interests would only take one's mind off work—the direst sin of all.

Walton said both Tom Tyler and Bill Ellison had unusual passion for their work. (They would both die a premature death.) "That extreme passion would sometimes embarrass me as we might sit in a quiet restaurant and he would start to pound on the table over some point he was making about something that happened that day," Walton said.

"Once the Bull told me that sometimes when he traveled with the bosses from back at the mill or various wood dealers and they said something that would upset him, he felt like he could tear out the steering wheel from the dashboard," Walton said. Now that's passion.

Walton remembered a special time when Jim Vadas tried to help Wayne Beasley and him regarding getting some time off during his first Christmas with Chesapeake. "Wayne was from Georgia and I was from New Jersey and the official three days off at Christmas the Company was giving us meant neither of us would be getting home for the holidays," Walton said.

At a staff meeting Vadas adjusted his glasses a few times, cleared his throat, then stepped up to bat for the new employees. "Bill," Vadas began carefully, "Christmas falls on Wednesday next week and since we only have three days off, that means our new men from

Georgia and New Jersey won't be getting home this year for the holidays unless we can arrange for them to get at least one more day off for travel time."

There was absolute silence in the room. Walton remembered how terrible it was as they sat there holding their breaths and squirming in their chairs awaiting the verdict. Finally, after what seemed like an interminable length of time, the Bull began to speak slowly. "Well, Jim, you can tell the BOYS," followed by a long pause as he cast a most distasteful glance in Walton and Beasley's direction, "that they have my utmost sympathy!" End of discussion.

Very few men other than Vadas would have risked so much for the new "boys" on the street but he was the sort who would always stick out his neck for those who worked for him. But what Vadas did at the staff meeting turned out to be of some avail after all.

Later, Walton remembered, when Vadas was preparing to leave for a week in Ohio and he and Wayne had dismally made plans for their big "holidays in Henderson" extravaganza, they were suddenly called into the Bull's most holy presence.

"What do you guys have planned for next week?" Ellison asked.

"Wayne mumbled something about going over to Eno woodyard to check on things and I mentioned Butner and Gulf woodyards," Walton said.

"That's not so important that it can't wait," the Bull responded with a smile.

"It was almost scary, that smile," Walton remembered. "Wayne and I, expecting a trap, reiterated our joint need to check on these woodyards just as soon as possible."

That's when it happened. The heavens lit up as the Christmas spirit hit Ellison. Then the great voice spoke. "That can wait," said the Bull. "Why don't you two guys take the entire week off so you can both go home for the holidays?"

Walton said that in spite of himself a lot of the Bull must have rubbed off on him. "Just recently, as I was traveling with a brand new forester we had just hired, I caught myself repeating some of the 'Bull's Bill of Rights!' "

The Bull may have passed on to his great reward much too early but he left a big part of himself back on earth with the Company he lived and died for.

The Bull passed on some good management skills, too, or perhaps Walton picked them up in spite of his exposure to Ellison. Whatever, Larry Walton took his place as Henderson, North Carolina, area manager before moving on to manage the Eastern Shore after Bob Owens retired in 1989. Jim Vadas went on to become operations manager for Woodlands. And who's to say Jack King, himself, didn't pick up all of the Bull's management skills, even from Keysville? That may have propelled him on to chief of Woodlands.

Not bad results for the "Bull's School of Wood Procurement." Indeed, Bill Ellison would have been proud.

Chapter 13

The Harper-Miller Years

By 1985, the years of rapid growth, careful planning and leadership of Tom Harris were drawing to a close. At the very same time, Sture Olsson's years as chairman of the Company were coming to an end. Harris had spent 18 years as head of Woodlands with a close relationship to Olsson. With Olsson's total support, he had brought about many changes and a period of expanding national reputation for Chesapeake Woodlands.

As Harris was enjoying many well-earned retirement celebrations with his co-workers and friends, he left the Company with a meticulously developed business organization plan for the Woodlands department. He had carefully put together this plan to ensure that just the right people were in just the right places to carry on in the Harris tradition. Chesapeake had an uncanny talent for knowing how to use every person on the staff in just the right spot. Under Harris, this gift had been especially utilized.

The organization plan Harris developed for Woodlands in April 1985, just before retiring, had Harris as vice president in charge of Woodlands and Paul Harper as operations manager with Ida Dawson as secretary to both men. Already Woodlands had its own group financial analyst, John Wenrich, and safety director, Jimmy Sears and a human resource department headed up by John Hockman, assisted by Mary C. Bourne.

Woodlands manager was Sharon Miller, Wood Products was headed by Ron Roberts, and Delmarva Properties president was Dick Brake.

Under Miller, the name of Ida Dawson once again showed up. She may have thought at that time that she was carrying more foresters on her keyboard than the woods had trees. But Ida was one of those special women who had that rare capacity to become sort of a mother hen to her flock of constantly pecking chicks.

Such a person can actually make or break a business. If Dawson had too much to do all those long years with Chesapeake, she never knew it. She loved her flock of foresters and they absolutely adored her. Dawson solved more problems each day behind the scenes than one could possibly imagine. Her deep, luxurious and upbeat voice over the telephone representing Woodlands for so many years brightened the day for many a far-flung employee checking back with the home office.

Miller had many servings of responsibilities on his platter. He had Jim Willis, technical land services manager, Tommy Callis, marine superintendent, Jack King, operations manager, Walter Zingelmann, project forester, and even the comptroller, Jonathon Edwards, reporting to him.

On top of that, the five area managers worked directly under his office through Jack King. They were Area 1, which consisted of West Point and all surrounding counties between the James and Rappahannock rivers to Richmond, Bill Rilee; Area 2, Eastern shore from Sussex County, Delaware, down to the tip of the Virginia peninsula, Bob Owens; Area 3, Keysville and the Southern Piedmont counties north and west of Area 1, Jim Vadas; Area 4, area east of the Rappahannock River, Marc St. John; and Area 6, Henderson, North Carolina, and Virginia areas of Danville, Roanoke and south of Lynchburg, plus old Area 5, Larry Walton. (Area 5, which once had included the Elizabeth City, North Carolina, area and had now been merged into Area 6.)

The new organization plan called for each area manager to be set up with its own secretary and, in addition, had its own land manager and procurement supervisors with assigned foresters and technicians reporting to them. If the area, like Eastern Shore, had a chip mill in it, its area manager also had that responsibility, as did Bobby Owens who had a foreman and crew under his wing to run the operation.

In the Wood Products Division, headed by Ron Roberts, every profit center was headed up by a different manager. At the time, the manager of hardwood lumber sawmills was William Phipps, manager of Treated Lumber was Rae Ehlen, manager of Chesapeake Plywood was L.R. Duncan, and manager of Dejarnette Lumber Company was Douglas Jones.

The organization was perfect. Each person positioned in Chesapeake Forest Products had a special gift to give to the Company, exactly matched to where such a gift was needed. Harris had seen to it that each individual was in position to make sure Chesapeake Forest Products always met its economic goals.

With such a plan in force, at last Harris felt he could retire with ease after so long carrying the weight of bringing in the wood. The man chosen to succeed Harris was none other than his operations manager, Paul Harper.

Paul Harper was already in place and ready for the job. Harris had always fondly called Harper "One of my three stars," the other two being foresters Sharon Miller and Dick Brake. On top of his star quality, Harper was a man who worked hard but was also a barrel of fun.

Harper hailed from South Carolina, and, like his former boss, had earned his degree in forestry at North Carolina State. After graduation, he had spent 17 years at International Paper Co. and another several years as assistant woodlands manager at Great Northern Paper Co. He had been hand picked by Harris in 1968 to take over Chesapeake's number two position and was well prepared for his duties.

As Harris had been, Harper was well seasoned by years of experience in the paper products industry. He came to Woodlands with a full understanding of what the competition was up to. This inner industrial knowledge was of the utmost importance to Chesapeake. Harris knew enough to ensure with his new hire that the tradition of always knowing what the competitors were doing would continue at Woodlands.

Harper came to Chesapeake shortly after Harris had taken over as vice president in 1967. By that point the Woodlands Division had grown so big it had to split responsibilities between the vice president position and actual operations.

Harper remembered how he had come to Chesapeake. He had been at Great Northern when he suddenly got a call from his old friend, Tom Harris, who had just left Albemarle Paper Co. and gone with Chesapeake.

"Hop on a plane and meet me in Atlanta," Harris had told him. Before Harper knew it, he had been made the new Woodlands operations manager for Chesapeake and all six ter-

At Sheldon's Restaurant and Motel in Keysville in 1990 are, from left, Paul Harper, Tom Harris, Red Highland, Lealon G. Vassar and Lealon Morris Vassar.

ritories were now reporting to him.

"Right off the bat, Woodlands was split off from the mother company and we formed into two separate groups," Harper said. "We both had our own budgets, personnel and accounting departments."

When 1985 came and Harris retired, Harper was elected to the post of "group vice president," as Harris had been. As it turned out, Harper was the last Chesapeake vice president of Woodlands. When Carter Fox reorganized the Woodlands Department in 1980, the division once again went under the paper mill management. Harper inherited the same problems Harris had dealt with. Chesapeake Woodlands had not kept up with the rest of the industry.

Harris had to really push to grow Woodlands organization to the point that it could keep up with the demands of the mill for daily quotas of wood. When Harper took over he found, in spite of Harris' long years of hard efforts, Woodlands was still at least "six or seven years behind."

Mechanizing the process for getting in the wood was the best solution. This meant converting their long-term, hardworking wood dealers, wood producers and loggers into fully

112

modernized wood gatherers who used the best technology and state-of-the-art machinery and highly trained workers. Gone were the days when Chesapeake could rely on the wood gathering systems of yesteryear that were too slow, costly and unproductive. Today's wood dealers had to use skidders and chippers, and had to hire people who were highly skilled and able to use the equipment.

"It was quite a chore," Harper remembered, and a task that almost had to be approached on a man-to-man basis with Chesapeake, in many cases, having to actually do the financing on the heavy equipment. It had to be Chesapeake because the equipment was just too expensive for individual wood dealers and contract loggers to handle.

Harper said the number one question he and his staff and private contractors faced every day was how were they going to get wood for the mill. The question was of such importance that nothing else came before it. If the mill had no wood on any given day, it had to close down. So, like almost everything else, it boiled down to jobs. No wood, no jobs. Keeping the mill well fed with wood was a bit like tossing cookies to a constantly starving, bellowing Cookie Monster.

Harper grinned, looking like a boy with his hand in the cookie jar, when he remembered those hard pressed days. "Wood procurement is always a problem. It's a feast or famine situation like almost everything else in life," he said. "We did what we had to do to get the wood, and sometimes the way we chose wasn't the best way, but we always got the wood."

As emphasized previously, the only way Chesapeake could get the wood, even in famine times, was by putting the right people at the right place all through its system. Chesapeake had a knack for finding and keeping just the right team players.

During Harper's time they had five major wood shortages. The shortages were caused by constantly fluctuating market conditions or weather extremes. "Each lean period we endured, we had to sit down back in West Point and think of ways that we could solve the shortage," he said.

If the wood shortage were merely a regional problem, Harper could simply resort to doubling up quotas from other regions until the shortage difficulty subsided. Sometimes they could use wood from Company land reserves.

If it were a transportation problem simply changing truck sizes suited for wood lengths from five feet to full tree lengths helped rectify the supply problem.

In 1977, Woodlands suffered an unusually frigid winter and the bay completely froze over. The tugs could not move and the trucks could not get into the woods. Woodlands solved the problems by getting its wood dealers to immediately arrange for trucks usually used for hauling farm produce and grain to bring in the wood chips. Word spread rapidly that the West Point mill was almost out of wood. Almost miraculously, farm tractor-trailers loaded with chips started arriving at the mill from all over the three-state area. Because of that gargantuan effort, the mill stayed open.

Harper remembered that in 1979 they solved the wood shortage by using additional railroad cars bringing in the wood along with the trucks. A simple doubling up on the method of transport solved the dilemma.

But when the shortage of 1982 hit, it was the most severe shortage of all because it was south-wide in scope. Harper smiled at the memory. "We solved that one, too, because Harris' sustained wood plan started back in 1968 had finally prepared us with some of our own wood from our pine plantations to harvest. We were so well-organized for that shortage, we not only had enough wood for our own needs but we could also sell our own wood to other mills and cut a nice profit, too."

Harper was proud to say that in his years as operations manager the mill never once shut down completely for lack of wood. (They did have slowdowns, however.) Even when other paper mills had closed temporarily because of lack of wood, the West Point mill stayed open. The good times continued, thanks to the tree plantings of yesteryear that at last had started kicking in. By the mid 1980s, Woodlands was well on its way to building great reserves in timber. These reserves lent a layer of comfort to a Woodlands Department that had never before known any insurance pertaining to getting wood other than their own daily devotion to hard work along with a measure of old-fashioned wile.

"The worst thing any woodlands department can do," Harper said, "is bury its head in the sand as to what other competitor woodlands departments are doing. What a woodlands group is doing down in Georgia will eventually affect us in Virginia. So we have to know about everything that is happening in the overall industry."

Eastern Shore supplies were so important to the mill, Harper remembered, that when Bobby Owens suffered a heart attack, he and Tom Harris were on a plane immediately flying across the bay to see him in the hospital. Harper knew the heart attack had probably been caused by the high stress of providing daily wood quotas.

Later, Paul Harper would suffer two of his own heart attacks. "Up until that point, I took my briefcase home with me every night, and nighttime work was just like daytime," Harper said. The doctor told him no more work at home. Harper followed his advice for three weeks and then returned to his old behavior. It is almost impossible to teach an old dog a new trick.

When his doctor heard about his return to old habits, however (and in a town the size of West Point the doctor knows everything), he called Harper up and said, "Paul, you don't need an appointment with me, you need an appointment with your undertaker." That remark finally got the old dog's attention.

Another big problem for Harper was the Chesapeake Bay Plywood Company over on the Eastern Shore. The Company had put in a chip processing plant and plywood log merchandising operation right next to the plywood company but they had problems with this situation right from the start.

"The plant could not meet the demands for plywood logs to the plywood mill," Harper remembered. His main challenge was to try to speed up the process. Bob Owens was an indispensable support for Chesapeake in all its Eastern Shore challenges.

Harper was one of those rare people who had expertise in his profession but also was blessed with a vibrant personality. He never lost his sense of practicality, in spite of his academic background, or his robust sense of humor.

"I remember a college professor back at North Carolina State once told me, 'Now that you've learned all about the trees in the forest, Paul, let me tell you about a dollar.'" Harper laughed at the memory. The professor continued giving his students good advice: "The minute the trees in the woods don't look like dollar bills, it's time to get out of the wood business!"

Companies that work hard usually form highly cohesive work groups. This triggers close bonding, friendship and fun. Harper had a lot of fun over the years with his closely knit staff.

"Once Tom Harris promised Bob Owens he would take him on a trip to New York City if we could lick this latest wood shortage," Harper remembered. "Owens told Harris he preferred a nice fishing trip. Finally the lean times passed and Harris arranged for the payoff with a fishing trip on Eastern Shore. The only problem was the seas were rough that day. Harris was soon as white as a sheet."

Harper laughed again at the memory. "Bob noted his boss's sickly white appearance and said, 'Now, Tom, isn't this nice fishing trip a whole lot better than a trip to New York City?'"

Jack King and Paul Harper hunting in North Carolina during the early 1980's.

Harper also remembered the first general Woodlands meeting Harris held for his personnel. Harris took one look at his raucous staff and decided he needed to set a few rules really fast and right up front.

"There will be no drinking after eating!" Harris barked to the not so attentive group. Vic Kaczmarski turned to Harper and said in a voice that could be heard way across the room, "But he didn't say what time we had to eat, did he!"

Harper also remembered the beloved character of forester Bud Johnson. The men took many trips to Eastern Shore together for good duck or quail hunting or fishing. Occasionally sleeping in a little later than Johnson wanted them to led to lots of good fun.

"Bud always liked to go fishing before daylight," Harper said. "He would always try to hurry us out of bed and into the boat and out to the bay and then drop anchor to fish. But every time we got out there and were set to fish, Bud would say, 'Well, I guess we better wait a while before we drop our hooks for the tide to change!'"

"Hard at work," Joel Cathey fishes with Paul Harper on the James River in the late 1980's.

Jimmy Sears was full of fun, too. Harper remembered one night when he and Sears were in Birmingham, Alabama, on business. After work the first night, they decided to go across the street to a bar and have a little drink. "As we approached the door, the doorman turned his head inside and announced in a loud voice . . . 'Get ready girls, here come two LIVE ones!'"

But the favorite story of all took place when Harper, Sharon Miller, Steve Boynton, who was Chesapeake's chief engineer, and Jack King were on a trip in Mississippi. The hosts had taken them to their hunting lodge some ten miles into swamp country where they had drinks and dinner. The plan was that after their hosts left, the Chesapeake team was to spend the night on their own.

"Around 10 p.m. we finally realized all our hosts had gone home and left us without a single bit of transportation," Harper laughed. " 'If one of us gets sick in the night, someone will have a mighty long walk in the dark through this swamp to get help,' I advised the suddenly very somber group."

Everyone was quiet for a minute while the seriousness of the predicament set in. Then Jack King remarked, "I sure hope I'm the one who gets sick!"

Harper loved his work and the men and women at Chesapeake. When he retired, he enjoyed the big celebrations in his honor, too. "They had many parties for me," Harper said with a smile. "Woodlands, wood dealers, even the loggers came, too." It only proved what everyone knew about the wood business and Chesapeake. The people worked hard but they also knew how to have a good time.

The second man to be called by Harris soon after arriving at Chesapeake in 1968 had been Sharon Miller who, at the time, was on the faculty of the State University of New York College of Forestry at Syracuse University. At that time Harris already had in place his operations manager, Paul Harper. He next needed his chief forester.

Harris had previously met Miller, who had once been over at Union Camp Paper Co. across the river at Franklin, Virginia. He knew of Miller's background of strong academics and practical experience. Miller would be just the man to help boost the blitzkrieg needed to bring Woodlands up to industrial par.

Of all the Chesapeake foresters, probably Sharon Miller had the most impeccable academic background. He was certainly the most published forester, for throughout his career he had over 33 articles and research notes published in periodicals within his field. He provided Woodlands with professional recognition from its peers when he first came to the Company in 1969 as chief forester.

Miller certainly gave Chesapeake a great deal of national prestige because of his ability to get his papers on the subject of forestry published in a multitude of national trade journals. He also held many honorary posts within his profession. These included service as a member of the National Council of the Society of American Foresters, committee member of the Southern Forest Institute, member of the Board of Directors of the Virginia Forestry Association and Forest Farmer's Association, along with many other organizations in the industry.

Miller went right to work assuming full management of Company forest land. To Miller, this meant developing a "formal" inventory of the extensive Company holdings and putting this inventory on a computerized record system for the very first time. "We started at point zero to develop our new plan," Miller remembered.

The new "sustained yield management system" that Harris was putting into effect at that time to ensure future timber to meet Company needs used an exceedingly simple formula for success. The same number of acres that were harvested each year would then be replanted. It was that easy.

But the system needed time to build mature timberlands. That was the one great flaw in the new program. When the Company needed extra cash a few years down the road, it soon learned to raid the timberlands. This enraged some foresters and rightly so because it sabotaged the program which had been so carefully planned.

Miller proved the value of the Company's occasionally having a forester out of the woods. He sat on many national forestry boards, served as chairman on some and received many distinguished awards for his work in forestry. This gave the Woodlands department and the mother company prestige among other paper manufacturing companies.

There was always a lot of fun going on in spite of the hard work and pressure. Forester Jim Willis remembered once when Miller, Steve Simmons and he were in Blacksburg and had flagged a taxi to take them to the forestry school at Virginia Tech. Miller, who was particular about details, noticed that the cabbie's gas gauge was on "E." "Are you sure you have enough gas?" Miller asked the man. "Yeah, man, no problem!" came the optimistic response.

As could be expected, they ran out of gas and soon came to a sputtering stop. The cabbie turned to the men and asked, "Hey, man, can you guys get out of the car and push me down to that gas station way down the road?" Willis remembered that Sharon Miller led the way without batting an eye and soon three Chesapeake foresters dressed in suits and ties were pushing the taxi and driver down the street. Needless to say, the cab driver did NOT get a tip that day for his ride.

After Tom Harris retired, a big change took place in Woodlands ordered by CEO Carter Fox. Each department throughout the Company was made into its own "profit center," which included a budget and full responsibility for meeting it. Not everyone was overjoyed at the news of the big change.

When Tom Harris heard of the change, perhaps at one of his Wednesday morning breakfast meetings where the retired "old boys" met to continue their long and close friendships and to trade bits of news from the Company, he frowned. He knew that short-term profits and the long-term goals of Woodlands were not always compatible.

This was because long-term goals meant planting trees and letting them grow undisturbed for future valuable saw timber that would generate cash flow for the Company in future years. This was in direct conflict with the Company's grabbing immediate profits by cutting immature trees before they were ready.

Harris dourly predicted the end of Woodlands to his fellow retirees. How true that prediction came to be.

By 1989, Company pines were only two-thirds the age of their normal 35 year maturation. But even at such young age, they could still be harvested and used for pulpwood. This reserve, as young as it was, gave the Woodlands Division a bit of a cushion to deal with lean times.

But the continued need to generate cash flow caused the Company to move in on immature forests on a regular basis. "It was frustrating to see the sustained yield system we had so carefully devised continuously altered for short-term fiscal purposes," Miller remembered. "But at least the mill had enough pulpwood to keep going even in short periods."

By 1987, Miller had been promoted to director of Woodlands and Wood Products. This change was a direct result of the Carter Fox mandate that each department would be responsible for its own profits and expenses. It was a huge shift for Woodlands to suddenly be thinking of wood products they could sell for a profit other than simply growing and harvesting trees to keep the mill in good supply.

One of the most difficult challenges Miller had to face was Chesapeake Bay Plywood Company on Eastern Shore. "We did everything we could to keep it open including trying to deal with the woodworker's union," Miller said. But profits continued to worsen, even under the able direction of Wood Products Manager Ron Roberts. A strike over wage negotiations finally ended the venture and Chesapeake claimed that the out-of-town union leaders who had stubbornly refused to accept the only deal possible had been to blame.

Miller said his biggest disappointment with the Company was the change in philosophy. "We took a short-term view rather than maintaining a long-range perspective," Miller said. This is a mentality that is quite foreign to a forester's nature. "In order to meet cash flow demands each month and a profit each quarter, we had to cut younger and younger trees.

"But Woodlands does not operate like this," Miller explained. "Cutting younger and younger stands of pines might help the cash flow temporarily but it is a false economy. If we had waited even five to eight more years, it would have made a big difference in output from the sustained yield plan, therefore generating more profits in the future."

But Carter Fox was CEO and times had definitely changed. One problem for Woodlands was realized by retired Tom Harris. "No longer did the Woodlands chief have quick and easy access to his CEO," he said. "In my time, I could call up Lawrence Camp or Sture Olsson anytime I wanted to and get an instant approval of any plan I had," Tom Harris remembered. "But now there was a new layer of management between managers and departments, which slowed management decisions to a snail's pace."

Miller said it was heartbreaking to see the carefully laid plans that Harris, Harper and he had developed over the years almost entirely junked. "Our young forests were raided anytime they needed money," he said sadly, shaking his head. A forester understands such sentiment. It was almost as if young girls were asked to go out with lecherous old men before they were even ready to dance.

Miller retired as director of Woodlands and Wood Products in 1991, offering only short respite from the North Carolina gang that seemed ever ready to rule Woodlands. He would be replaced by yet another North Carolina State forester graduate, Jack King.

Chapter 14

The Foresters

What is a forester? I have a distinct memory of asking myself that very question when Jim Vadas, identifying himself as a "forester at Chesapeake Woodlands Division," had originally called back in 1996 to speak with me about the possibility of writing this book. What in the world is a forester?

In my mind's eye I pictured some strange fellow dressed in a quaint, elfish costume, on the order of Johnny Appleseed, traipsing through the woods with a sack strapped to his back filled with baby trees. Every so often this odd guy would stop and plant a tree. He might do other things in the woods, too.

He might take out a stick, measure a few trees, eat some fruit or nuts, maybe kick a few stumps, perhaps even throw a couple of acorns at a chattering squirrel, then move on to the next important assignment, whatever that might be. Who really knew what a forester was up to with his special academic qualifications once he got into the woods?

Of course, I did not tell Vadas that that was what I imagined he and all the other foresters at Chesapeake were up to. Writers do have a measure of sense, yet God only knows what we do all day long sitting in front of our computers in the act of writing a book.

I decided to take out my dictionary and look up "forester." American Heritage supplied three definitions: 1. a person trained in forestry, 2. one who inhabits a forest, and 3. a brightly colored tropical moth. I felt pretty sure that the last definition did not apply when it came to Chesapeake foresters because in over 75 interviews, not one of them had ever displayed a colorful wing.

Still, my first session with Vadas was a real eye-opener. I immediately sized foresters up. I decided foresters were darned strange. Even though this judgment does come from a member of the writing profession which can be known in certain groups for a certain eccentricity, it was still my hallowed opinion.

I'll tell you why. Vadas and I were driving to Keysville, Virginia, speeding along Route 15, talking about the book and our plans to meet and interview the wood dealers, when I happened to say (quite innocently, I thought), "I wonder what that purple-flowered tree is that we just passed on the side of the road."

My mistake. Never ask a forester the name of some passing tree unless you really want to know and you have a lot of time on your hands to go back and check the tree out. In a flash, Vadas pulled off the highway, turned around, and we were back parked in front of the purple-flowered tree.

Vadas jumped out of the car with the same enthusiasm I give to a book of poems. He then proceeded to approach the tree in a manner that looked something like a primitive, tribal chieftain beginning some odd ritual with the intent of worshiping a tree.

He had a certain manner about him, as if he were in the presence of some really holy deity. He walked carefully, quietly over to the tree with a good measure of respect, as if he did not want to offend the tree. Maybe he asked the tree if he could approach but I did not hear it since I was sitting back in the car. Who knows? But soon he had apparently bonded with the tree and was fiddling with it like I suspect all foresters do with trees.

Pictured around 1980 are, from left, Jim Vadas, Jack King, Bob Owens, Marc St. John, Paul Harper, Bill Rilee and Bill Ellison.

First he broke off a twig and sniffed it. Then he tasted it. Then he managed to latch on to one of the purple blooms and did the same weird stuff with that. Then he got up really close to the tree and examined the bark, sort of the way Mr. McGoo would examine the pages of a book, running his hands over it, thinking about how the bark felt, how many ridges or nubs to a finger, the whole bit. Then he stepped back and took a good long gander, maybe checking how many insects were dwelling in the bark, taking in every detail of God knows what—all this just because some fool wondered about some tree.

I pass thousands of trees each day, maybe even millions of trees, and I might wonder what they are but I am somehow able to get through my life without knowing each tree's exact name. I have more pressing things to do, such as studying Keats or Joyce, but not Vadas. I saw the look in his eye.

I knew that look. I had seen it before many times in life. I called it "determined zealotry." And I knew we would not budge one more mile toward Keysville until this tree was identified.

I was about to wonder if the third definition of a forester was correct, if maybe this fellow weren't a brightly colored tropical moth, after all. To say I stared in rapt astonishment at Vadas as he inspected the tree would perhaps be an exaggeration. But I had to admit I had never before seen a real, live human being behaving in such a strange manner in front of a tree. It got my attention. From that point on I held a careful respect for foresters and never again asked the name of any passing tree.

Still, I have to admit for purposes of this book that if one spent a lot of time with a forester, one might learn an awful lot about trees and wildlife. The fact is, foresters study a special curriculum in forestry schools that is every bit as challenging as any other science.

At one time someone like Tom Tyler who worked for Chesapeake from the Eastern Shore area could rise to forester and even receive a forestry license just from experience working in the woods and without attending a forestry college.

But those days have long passed.

The number two Chesapeake forester to be hired by Woodlands, Ed Tokarz, remembered starting work as a brand new forester for Chesapeake in 1940 just out of Virginia Tech's forestry program for $20.00 a week! That, too, has changed. A starting forester just out of school can now expect to begin work in the range of $25,000 to $30,000 per year. (That, too, has changed.)

Forester Walt Tilghman uses an "increment borer" to determine the age and growth of a tree.

How else to define the work of a forester? A special soul is the forester, half science, half poetry. He or she both loves the forest and all that is in it and wants to protect it, but also has a healthy understanding that civilization must have the use of wood products in order to exist.

The forester is never one to go berserk over the beauty of trees. He or she knows that since the beginning of life many species have required the tree for food and shelter. It will always be. The noble trees were ordered in the natural scheme of things to carry much life upon their sturdy limbs.

The question then for the forester is how to balance the needs of mankind with the needs of other wildlife and how to ensure that the harvesting of trees is carefully replaced by periodic planned new growth, which protects all concerns for the future. The forester therefore is involved in duty, work and responsibility on both sides of the coin. He serves two masters, rapidly expanding society demanding more and more wood and the need for conservation to preserve the natural balance on earth.

A walk through one of Chesapeake's forests explains private sector forestry to the public better than a thousand words. Chesapeakes' woods are professionally managed and groomed, planted in superior pine seedlings, and beautifully maintained. Compare these to the forests at a national park where trees and shrubs are left in a natural condition, where

Forester Ed Tokarz presents a "Tree Farmer" jacket to Michael Hackney for his winning report on forestry.

the only management they receive is fires that burn them to the ground. This comparison makes the difference in natural and professional land management very clear.

Over the years Chesapeake has employed more foresters than any other private employer in the state of Virginia. Thus, Chesapeake can speak with authority on the subject of foresters and forestry. The following stories are about some of the foresters Chesapeake has hired over the years and the experiences they have enjoyed in and around Virginia.

At one time there may have been such a thing as a "North Carolina State Mafia" at Chesapeake Woodlands as one employee had jokingly mentioned (Harris, Harper, King, Brake and Ellison were all North Carolina State University graduates, three of whom were chiefs of Woodlands), but the Company hired foresters from a variety of forestry colleges. This practice kept the Woodlands department from becoming too insular or provincial. (Not that North Carolina State could ever be considered in such a category.)

Indeed, Bud Johnson, one of the very early foresters, was a "Yalie." A great deal of credit must be given to the person who was responsible for letting the Ivy League into Chesa-

122

Edmund Tokarz and wood dealer Bill Sanders with one of the first Chesapeake planes used to visit the Pittmans Cove Barge Landing in the mid-1960's.

peake back in the early 1940s when it was a teeming bag of Southerners.

Virginia Tech's School of Forestry was represented at Chesapeake early on with the hiring of Ed Tokarz, who started with the Company in 1940. Tokarz can even go further back than that in the Company. He remembered spending a summer in 1938 as a college boy running chemistry tests on soil samples for Henry Vranian, chief chemist and subsequently vice president of sales and one of the early members of Chesapeake's Board of Directors.

Tokarz remembered starting work in Woodlands in 1940 for $20 a week. One of his first tasks was helping pull the cross saws that brought down the trees on the old Hickory Hill tract in Gloucester County, Virginia. For a few years, he and Gooch were the only foresters in Woodlands.

Tokarz is one of the few foresters still living who used the old crosscut saws at Chesapeake. "In the mid 1940s, the two-man power saw came in and the old crosscut saws went out," he said. "So then we had a new problem. The men kept forgetting to mix oil in the gas and the engine burned up and had to be sent back to the shop!"

Tokarz remembered telling Gooch one day that $20 a week was not enough pay for a college graduate. "You're not married and you don't have a home so what do you need money for?" Gooch responded. A year later Gooch raised him to $21.85 a week. One can only imagine how Tokarz must have felt to be suddenly cast in such high cotton.

The war came and Tokarz had to go into the Army, but Gooch assured him his job would be waiting for him when he returned. Tokarz trained troops at Camp Edwards and later went over to Europe. He was a part of Patton's thrilling advance and final breakthrough into Germany.

After the war, Tokarz came back to Chesapeake and was assigned to the procurement office. He eventually became manager in procurement. He was area manager in West Point for several years. Then his career with Chesapeake landed him in Oxford, North Carolina, as a procurement forester. During those years Tokarz experienced great pleasure in his work by working with private farmers to improve their timberlands.

"The greatest moment for me in North Carolina was a 40th year anniversary celebration of a Tree Farm in Oxford," Tokarz said. The event was part of a program of the National Tree Farm Association which is part of the American Forest Institute (AFI). It captured a lot of attention in the media for Chesapeake and triggered a great deal of interest from local landowners as to how to improve their trees.

"We had woodyards fed by two railroads, Southern Railway and Norfolk and Western Railroad," Tokarz said. "We were on a quota to provide wood to West Point." Keeping those racks filled was a constant pressure for the Chesapeake team.

Tokarz certainly remembered the tough years down at the woodyards, loading the train car racks with wood. It was hard work, often done in the worst possible heat. Work crews needed constant, close supervision to make sure the racks were loaded properly.

Once, when he was 62 years old, he was up on a wood rack breaking in two new employees. While he was showing them how wood was loaded, the railroad rack car tipped over. Tokarz was almost killed in the accident. "I was in bed for six weeks recuperating," he recalled.

Tokarz remembered feeling that the North Carolina foresters stuck together and ran everything in those days. It was true that Harris and Harper were both from North Carolina State, but they were followed by Sharon Miller and he was a University of Michigan man. There might very well be something wrong with North Carolina, but what could possibly be said against dear ol' Michigan? Certainly this Ohio-born, Buckeye writer would never say one word against anyone from the beloved state of Michigan.

Although Chesapeake has certainly been somewhat dominated by leadership from North Carolina State foresters, Chesapeake did try, even from the beginning, to hire foresters from all over the country. At its peak Chesapeake Woodlands had up to 50 foresters on staff with over 15 schools of forestry represented.

Tokarz retired in 1983 after giving 40 years of service to the Company. His interview took this book directly back to the number two forester hired in Woodlands.

Forester Fayette Wiatt, a native of Gloucester County, Virginia, who came to the Company in 1965, was involved from the start with intensive land management. The very first job he did for Woodlands was burn a 564-acre company tract. "I never will forget it, either," Wiatt said. "In forestry college, we practiced our burns on a three-acre tract and what a difference an extra 561 acres can make!"

A "prescribed burn" is necessary to remove debris and to control competition with new seedlings. At that time, Wiatt remembered, a pine seedling was costing the Company about a penny a tree, and lost or unplanted seedlings could quickly build up to considerable costs.

"The workers were paid not by the hour but by the number of pine seedlings they planted, so we had to keep a firm eye on what was going on.

"We would plant 800 seedlings per acre and we had to make sure those seedlings were actually planted," he said. Wiatt learned to check the work closely. It was not unusual to find a clump of seedlings thrown in the brush. "Once I found 600 seedlings thrown in the woods and another time I could not find 30,000 of our seedlings, so I went back to check the plantings and found 30 or 40 seedlings had been buried in each planting hole!"

One day he was checking a Company tract where pine seedlings were not coming up well and he came up on a group of five workers standing in a gully. "For some reason I was immediately suspicious," Wiatt remembered. "I walked right over to a fresh burial site and unearthed a big cache of buried seedlings."

This kind of situation was not unusual in the forest management business. Even crews who were paid by the hour were burying trees since they had daily quotas to fill and this left the men free to slow down their work. "In all my days I never knew a planting crew who did not bury seedlings," Wiatt said.

Wiatt was eventually transferred to the Fredericksburg, Virginia area to manage some 20,000 acres of Company land. He also was assigned to assist the Fogg Brothers, who were wood dealers in the area, and help keep them supplied with wood for their chipping crew.

Wherever the Company had land or worked with wood dealers, a Company forester was

Which one is the marine superintendent? Hint: the one not wearing a tie. Tommy Callis claims the reason some decisions are made, or not made, is because a tie has cut off the blood supply to the brain! The Forest Products staff members shown are, from left (seated), Paul Harper and Jack King, (standing) Ron Roberts, Callis, Jim Vadas, Larry Walton, Steve Whaley, Fayette Wiatt and Marc St. John.

125

bound to be near at hand. Foresters had to be ready to solve any kind of problem in the territory where they were assigned. Over the years, the Company foresters saw almost everything.

"Once Dick Brake called me and said someone was stealing timber on a 2000-acre Company tract and for me to go check it out," Wiatt remembered. "But no matter what I did, I could never catch the fellow. I would go in one side and he would go out the other." Wiatt laughed at the memory of seeing his bobtail truck tracks in the snow but never quite getting there in time to nab him.

"I finally caught him and when I came up on him I told the man, 'Hey, I'm from Chesapeake and we need you to help us cut our wood and could we hire you to do that for us?'" The man agreed and Wiatt said that finally solved the problem. This story only illustrates the point that it sometimes takes a very clever plan to skin a very clever cat.

Another time when Wiatt was burning a 2000-acre tract in Nelson County, he found no less than six whiskey still sites on a stream that passed over Company property. "One day I found all six of these stills were up and running at once. The last man on the creek complained to me the creek water was too damn hot to run his still!"

One of the most interesting sights Wiatt saw during his years of running inventories for the Company was during a walk over on Hawkins Mountain in Nelson County in the western part of Virginia. He came across a maple tree growing on the side of the mountain that measured 6 inches in diameter (4 feet up from its base) and was 92 feet tall! He believes that is a record.

His years of service made him very close to Randolph and Manley Fogg in the Fredericksburg, Virginia, and Northern Neck area. Randolph Fogg especially used Wiatt as a trusted advisor. Once Fogg was stumped. He had a piece of property planted with too many trees "going in" (at least according to what his records showed he was paying for) but not enough trees "coming out" (or actually being planted)!

Wiatt immediately suspected the old problem of seedlings being buried. "I went down there and inspected and sure enough, I found buried trees. I decided to put a red flag up everywhere I found buried trees. Then I called the Virginia State Department of Forestry, who had lined up the crew to do the planting, and showed them all the red flags. It was some embarrassment for them but a problem solved for Fogg."

Wiatt said Fogg had bought over a million dollars worth of equipment and the Company wanted to do anything it could to help him make money and stay in business. "We foresters watched out for these loggers every way we could and tried to do all that was possible to help them.

"Once we had bought a cutting right on a piece of land and the Foggs were chipping the trees," Wiatt said. "When they were finished, I decided to walk over the entire property to make sure everything was right. I found the skidder had left eight loads of merchantable wood behind!"

Another time after the Foggs announced they were finished on a cut, Wiatt walked the entire property just to make sure. "I called Randolph that night and asked him to send me two skidders the following day. There was so much wood left we chipped five more loads that day and even had to return the next day to get three more loads."

These examples clearly show how important a forester's sharp, ever watchful eye looking out for the logger's interests could make the difference between profit or loss. What

Wiatt did to help Fogg Brothers was typical of going above and beyond the call of duty. But it also explains why the loggers and Chesapeake enjoyed so many years of close, profitable relations.

Jim Willis, who is now operation managers for the three present Chesapeake Woodlands' territories, came to the Company in 1971 with a master's degree in forestry from the University of Georgia. Willis started work as a technical services forester and took on various responsibilities until he replaced Walter Zinglemann in 1974 as a technical services supervisor.

Willis definitely saw a change in the wind for Woodlands when in 1978 he was assigned additional duties of site maintenance supervisor. Times had changed and more economical and effective land preparation methods could be implemented without the use of the expensive heavy equipment of yesteryear. Under his direction, the Forestry garage was downsized and finally closed down altogether in 1993, thus saving the Company a lot of money in overhead.

Chesapeake was sticking to its new plan. Under Carter Fox, every department in the Company was expected to make a profit and become a profit center in its own right. If it could not make a profit and stand on its own, it was closed down.

The garage was doomed, as were other departments in Woodlands, under the new rule of profit or perish. It was only a matter of time before it would be forced to close. "We were just too small to take in enough work outside of what we provided to the Company to ever become profitable," Willis explained. "And we were too big in overhead expenses for the Company to carry us if we did not." The bottom line won out as it must in any corporation. It was cheaper for the Company to hire out its big equipment work to private contractors.

Willis mentioned some of the new technology that came into use at Woodlands during these years. The new satellite mapping equipment allowed Woodlands to connect to an overhead satellite to get detailed information on any piece of land they needed. This was a great improvement over the aerial photography that had provided information to the company in earlier years.

Another innovation came with global positioning satellite guidance for spraying the timberland. "Our pilots could use these satellite maps on their own computer screens in the cockpits of their airplanes to direct their every move," Willis said. The new techniques offered great improvement in keeping exact records as to what areas had been sprayed and what areas still needed to be done.

Willis eventually became technical and land services manager, proving that the job of a forester can move into many fields. He was in charge of such responsibilities as forestry research, surveying, tree improvement program and overall land management.

"Forestry has evolved rapidly as a profession," Willis explained. "A forester has to not only be a forester today but he also has to be an expert in public relations. You have to take the time to explain to people what you do and why you do it. This is because of all the environmental interest today within the general public. The truth is, most people don't understand what foresters actually do."

Willis said the Company used to have two full-time foresters who worked totally in education and public relations. But those days are gone. The job of public relations has fallen on everyone who wears the Chesapeake Corporation logo. "We never have enough time to do public relations, but every one of us makes the time to make sure the public understands fully what we do and why we do it."

Like other managing foresters, Willis has spent a great deal of time interviewing other foresters as job applicants. "The Virginia Tech Co-op Program has been utilized within Woodlands for over 25 years and has provided us many fine interns. The advantage is obvious. When a job vacancy comes up, we have an automatic pool of excellent applicants who already know us and we know them."

This illustrates again how many changes have come to the profession of forestry at Chesapeake. Just how many of yesteryear's foresters were ever prepared to do job interviews as a part of their job?

"Our interviews are all board interviews with a panel of at least four Chesapeake people. We especially want to know attitude to work and how well the applicant can work as a member of a team," Willis said.

"We find out a lot about applicants by asking specific work-related questions like . . . 'What would you do if such and such happened on the job?' This gives us an opportunity to observe how this person thinks and reacts to problems and also a good idea whether this person would fit into our team approach."

Larry Walton is the present regional manager of the Eastern Shore area. A reserved, cerebral, and reflective man, he, along with Jim Vadas and Jim Willis, demonstrates a big change in management "type-casting" at Chesapeake.

The once bombastic, hard-charging dynamo of a manager who drove his people like a team of wild horses is out of style. The new, sensitive era has begun, perhaps in tune with the team approach to finding solutions to rising challenges in a "friendly, non-threatening and caring" corporate environment. Its number one goal is to bring every last employee into the once tight circle of figuring out the universal problem of how to make a profit.

Walton, a native of New Jersey who received his B.S. in forestry from Clemson University, is an example of how well the Chesapeake summer intern program is working for the Company in scouting out good future managers. He was one of those interns.

"In the summer of 1973, just before I graduated from college, I came to West Point," Walton said. "I got a room for $10 a week in Tappahannock and every day I joined about 15 other interns and took inventory on Chesapeake land holdings. This inventory has to be done every five years so it was a perfect way to bring interns into the forestry process," Walton said.

"Each of us walked through the woods with a clipboard with a grid on it for making notations. We stopped every now and then and took soil samples to make sure the nutrients were proper for the trees. That summer alone I feel like we must have walked the entire 320,000 acres of Chesapeake's timberland!"

On the Eastern Shore, Walton not only deals with all problems of Company owned land maintenance and reforestation programs, but he oversees production of saw logs for the sawmill in Princess Anne, Maryland. This sawmill alone produces 25 million board feet a year. Walton also sees to it that the Company chip mill at Pocomoke, Maryland, receives all the pulpwood it needs.

On top of that, Walton has to oversee more than 300 hunt clubs that lease Chesapeake land for hunting purposes. Leasing fees have been a big source of annual revenue for the Company now totaling over a million dollars per year.

Larry reflected on the big change in Chesapeake philosophy toward environmental "green groups" like the Chesapeake Bay Foundation and others. "Once we might have said something on the order of . . . 'You couldn't pay us to walk into that "communist" camp,'" Walton said with a chuckle. "And now we are the closest allies possible." This bonding was bound to happen. Each side needed the other and we finally realized it. We began to trust and work with each other.

Walton spoke of another change in Company policy. "We now have 'management teams' that consist of 10 to 20 people per team and everyone is brought into problem solving."

This approach, used across the board by most American corporations today, has evolved slowly in our highly individualistic society. Over the many years of making profits, management finally realized anybody in the Company, from the CEO down to the lowly janitor, might hit on a good solution. The fact is that sometimes the closer one is to the actual problem, the clearer the solution is.

Also, good ideas for new products don't always happen just at the top. Any employee can have the ability and vision to think up a new product and help the Company make big future profits. Companies today want to create the environment where employees can have free expression on new ideas and immediate access to top management.

Walton spoke of the new era of bringing in the wood. "Chain saws replaced the ax and crosscut saw of yesterday, and the new saws are much more user-friendly and actually start on cold mornings!" he said with a laugh.

Hydraulic directional shears now snip off the tree at its base and then trim branches off the tree. Skidders load the trimmed whole tree on specially designed trucks that are built to accommodate the entire length of the tree. Radial saws with round blades that saw instantly replaced the older slower saws. "Sometimes trees are even chipped right at the site where they are harvested by the chippers, then loaded on barges and sent directly to the mill," Walton added.

It makes one wonder what the future foresters at Chesapeake will experience with whatever new equipment of tomorrow should emerge.

Forester Ray Ehlen spent many years in Woodlands working to improve reforestation on Company lands. "We discovered two big improvements during the time I worked in Woodlands," he remembered. "One was to improve the drainage in much of our timberland. We did this by building open ditches in the lowest levels of the woods to take off any standing surface water. We found our pine did much better with good drainage," Ehlen said.

The second improvement was called the "containerized seedling program" where the pine seed was first planted in a small pointed container rather than the ground. When the seed had matured, it could be planted anywhere. This simple device freed foresters from their December through April planting season time restrictions because the containers could be planted at any time of the year.

Ehlen saw many improvements in reforestation during his years at Woodlands. "When I first came to Chesapeake in 1970, it was not unheard of to see one of our crews clearing a field by dragging a log behind a John Deere tractor." But the Company rapidly built a big fleet of heavy equipment that could do just about anything that could be done in the woods with the very finest state-of-the-art machinery.

Maintaining equipment, however, turned out to be more expensive than the Company had initially realized. Chesapeake eventually realized it would be cheaper to hire out work demanding heavy-duty equipment to private contractors. In 1994, Woodlands made a big jibe in the wind. "All our big equipment was sold that year and our garages are now all empty," Ehlen said.

Ehlen had an exciting and profitable experience as a forester with Chesapeake as he and others built a successful wood treating company within Woodlands during the 1980s. More will be described about this valiant effort in a later chapter.

Chapter 15

Area Number One

What used to be known as Area 1, the closest area to the mill at West Point and made up of surrounding land, was first managed by Ed Tokarz. When Tom Harris came to town, he reorganized the entire territory system. Tokarz headed to North Carolina to work in procurement and Stuart Buck became the next manager of Area 1.

Buck, a short, thin, wiry man, was the sort of fellow who, even in retirement, might have some sawdust in his pockets or a few wood chips in his shoes. He first came to Chesapeake from the sawmill business, hired by Red Highland in 1966 from his position as right hand man at the B. C. Westmoreland sawmill in King and Queen County, Virginia.

Buck was a Gloucester county native raised on the land that is now known as "Lisburne," a beautiful, historic estate. He was a natural for Chesapeake, having served so many years (since 1945) at the sawmill. "I knew everybody at Chesapeake," Buck said with a smile, not to mention everybody else in the wood business. One rule in the wood business according to Buck is, "It's always who do you know who's got the wood this week that counts."

Buck began his work at Woodlands in the land acquisition office as assistant to Tau Crute. They developed a good system that worked well for both the Chesapeake mill and the Westmoreland sawmill. Any timber that could not be used at the mill any given week was sent back to Westmoreland. That way, there was never any excess wood.

When Tom Harris appointed Buck to be manager in Area 1, the new territory was drawn to run from West Point to Gloucester Point, Virginia, and all the way up to St. Mary's, Maryland. "We had 93,000 acres of timberland in our area and I had control of everything in it," Buck remembered.

By 1974, Buck was promoted to chip procurement manager, a post at which he remained until he retired. "I knew all the sawmill people all over the Chesapeake area and each year I bought 20 percent of all the chips that the pulp mill required."

Buck was a perfect chip procurement man. Since the quality of wood chips was very important (a load of bad chips could shut down the machinery at the mill), Buck personally called on all his sawmills every two weeks to make sure of the quality of its chips.

"In those days Chesapeake did business with 30 sawmills so I was very busy with my visits," Buck said, with a crafty smile that almost dared anyone to unload a bad chip on him. One could imagine what it was like at those sawmills when Buck popped in for a little visit and either those chips were not ready or they were not quite up to Chesapeake par.

But by that time, Chesapeake had started chipping its own wood and only buying chips it could not produce itself. Slowly, as its own chip production grew, the Company had less of a need to purchase chips from the sawmills.

Buck used to go into the office every morning by 7 a.m. so he could line up his customers for the day over the telephone. "Tom Harris was always there and he would always come in to see me and find out all the information for the day," Buck laughed. In a business that depended on who you knew and what you knew, it was mighty hard for another wood company to beat the "Harris-Buck" combination.

The wood inventory pile in the background, empty woodracks at the mill, and the conveyer (foreground) to the wood chipper are shown sometime between 1960 and 1980.

Buck enjoyed the yearly Lumber Manufacturing Association's convention more than any other event. Lumber people can sometimes be rough customers. A room full of lumbermen could challenge the toughness of any other industry . . . steel workers, dockworkers, oil drillers, watermen, even the gold rush crowd that headed west in the last century, none of which are necessarily known to be much higher on the scale than a pack of wild wolves.

"It was a tough and rugged business," Buck laughed. "The general philosophy in the wood business was . . . , if you cut my throat, I'll be back to cut yours!"

Buck also enjoys the weekly retired "old boys" breakfasts each Wednesday morning at Ann's restaurant in Glenns, Virginia, where retired foresters still put away scrambled eggs, ham, pancakes and hot biscuits like in the old days. "Tom Harris still asks me for all the latest information on wood supplies," Buck said, proving that once the wood business gets in ones blood, it never really leaves.

After Buck took over chip procurement, he was replaced by forester Walter Zingelmann as manager for Area 1. An Auburn University and University of Heidelberg graduate forester in wood technology, Zingelmann was hired in 1966 by Red Highland after serving a stint in the Korean War and working at the Virginia Log Company in West Point.

The famous "Wednesday Morning Breakfast Club" included, from left, Eddie Bernoski, Claiborne Courtney, Dick Brake, Dick Cartright, John Hockman, Stuart Buck, Edmund Tokarz, Tau Crute and Tom Harris.

His first official duty for the Company was evaluating the lumber located on a brand new 13,000-acre Company land acquisition located in King and Queen County, Virginia. "It took me three months to complete the task," Zingelmann said with a laugh.

He served as chip procurement coordinator until 1969 and then as technical forester service supervisor working with Sharon Miller. Part of Zingelmann's responsibility was overseeing Chesapeake's part of the Cooperative Tree Improvement Program with the School of Forestry at North Carolina State University.

By 1972, Zingelmann was appointed manager for Area 1. He served in this position until 1979 when he moved over with Dick Brake and was involved in land acquisition and cooperative forestry. He worked in acquisition until 1983 when he assumed the duties of staff forester specializing in timber and land closings, including the very knotty job of working with lawyers.

"I remember once we decided we would show the world how foresters manage a 'perfect burn,'" Zingelmann said with a smile. "Sharon Miller had drawn up the first Company environmental guidelines and we decided to put the guidelines to work. We had our burn on Company land over in King and Queen county and a TV station even showed up to publicize the event. It gave Chesapeake a whole lot of good publicity at a time when we needed it."

In 1979, forester Bill Rilee was assigned the additional task of serving as manager for Area 1 along with his existing assignment of being manager for Area 4, which was west and north of the paper mill.

Chapter 16

The Enigma of Carter Fox

Whether electing a CEO in 1983 other than Carter Fox might have saved the mill at West Point from eventually being sold off from Chesapeake Corporation in 1997 for just under $500 million, is a question to which no one will ever know an answer. But rightly or wrongly, blame for the sale and loss of the West Point mill will always be laid at Fox's doorstep by thousands of Tidewater citizens. This is simply because he was at the Company helm when the sale took place.

Who was this Carter Fox who started as an accountant in 1963 in the Woodlands Division and rapidly advanced up the ladder of command to the eventual rung of CEO, President and Chairman of the Board of Chesapeake Corporation? He was an attractive, shrewd, and very capable businessman with an unusual ability to envision the future. He also credits his rapid rise within the Company to his knowledge of computers at a time when computers were just hitting the workplace. Knowledge is power, and those who have the knowledge will surely get the power.

It was to Woodlands' great pride that a young Carter Fox spent some time in their midst as an intern with the Company over summer and later as a full-time employee after his graduation from college. It was in Woodlands that Fox learned to know all the foresters on a first name basis and to understand their fundamental concerns. Fox had a healthy respect for foresters and understood how important they were to the Company he would one day manage.

Forester Ray Ehlen noticed that Fox especially liked to hire foresters who held an MBA degree. That was because foresters, who were so honest by nature, did not necessarily have the business background that was needed. Woodlands eventually started the practice of sending some of its foresters back to school to pick up the MBA degree or at least some additional business training.

Carter Fox

Born in Aylett, Virginia, in King William County, Fox came from a family with a long association in the sawmill business. Fox remembered that when he was a student at King William High School one of Chesapeake's early foresters, Forrest Patton, came one day to speak to his class about trees. "He even took us into the woods to identify them for us," Fox said.

That might have been Fox's very first encounter with a genuine forester. Whatever, it left a lifelong impression on him. It must have been a surprise to Patton to know he had reached a high school student in one of his student lectures who eventually would become the head of Chesapeake. One can't help but wonder what might have happened if Patton's lecture on trees had been less interesting, or if the school had invited a vet or dentist that day to talk to the students about dogs, pigs or teeth.

Graduating from Washington and Lee University in Lexington, Virginia, with a degree in physics and engineering, Fox then went on to the University of Virginia to pick up an MBA, all the while interning during summers at Chesapeake. Upon his graduation in 1963, he was hired by CEO Lawrence Camp and went to work as an accountant in Woodlands. He immediately began the gargantuan task of converting Chesapeake Corporation to the big wave of the future, computers.

He was a bright person, considered a "golden boy," who immediately received and held the attention of top management. "It was a time when the bean counters were taking over management in most businesses across the country," Sture Olsson would eventually state. But his special ability was noticed right off by everyone who met him, including wood statistician Billie McKeever, who turned to Mary Ann Fetterrolf the moment she met Fox and said, "Mark my words, some day that Fox is going to run this Company."

The wave of computer technology that was hitting the shore of American business certainly helped Fox, and many others like him entering business at that time, take control of management. Fox easily convinced Camp that computers were coming and no company would be able to survive without them. By 1964, Fox had the first IBM 1440-14K computer installed and operating and had all records changed over to the new data processing system.

As Carter had the know-how to operate the Company computers, his authority and power grew until it permeated the very highest offices of Chesapeake. By 1967, he had convinced management that Woodlands needed a complete overhaul and brand new leadership. That new leadership soon came with the arrival of Tom Harris.

From the beginning, Fox instigated new ideas that eventually launched him right into the presidency of the Company by 1980 and to CEO by 1983. Some of his ideas transformed the entire Chesapeake way of doing business from the ground floor up.

Fox's changes were to allow for "an open business environment" in which all employees were encouraged to voice their opinions about anything that had to do with the business. "In that way management heard all ideas," Fox said. "The first idea may be no good but the second idea may be brilliant; so no corporation in today's world can afford to shut down any of its people."

Another Fox idea was to reduce fear within the Company so all individuals were not afraid to take action, regardless of risk, even if such actions might occasionally be a mistake. This was a change that had to come about. "Our companies were spread out everywhere," Fox said. "Sometimes management did not see another Company official for weeks. They had to be able to stand on their own, be entrepreneurial in spirit, willing to take risks and able to talk openly about goals and objectives and how to meet them."

Still, Fox took the blame for selling out the mill at West Point. The outpouring of rage in West Point and surrounding counties was evident, especially within the old boys' circle. There are some who will never forget or forgive his doing it. "They sold out our very own Company," one employee lamented. On the day the sale was announced another longtime employee faxed Fox one furious word, "TRAITOR!" Old Tau Crute, who died in 1999, went even further with his, "We'll never forgive him for this. Never. And be sure to include this sentiment in the book, too."

But Dick Brake had a very different way to sum it up. "You have got to give credit where credit is due. At least Carter Fox did the stockholders a big favor," Brake said. "Chesapeake did better after that sale than did the general industry. And after all is said and done, it's the bottom line that is the most important figure."

Trouble is, in business the bottom line changes with every quarter. Only time will be the final judge.

In spite of the human toll and obvious suffering within Chesapeake ranks, maybe the sale of the West Point mill was inevitable. Maybe nobody could have held on to it. Maybe changing markets and rising costs were more responsible for the loss of the paper mill than even Carter Fox was.

Some say the death knoll for the West Point mill began to toll in 1987, the year Fox moved the Corporate Headquarters to Richmond. A move that Sture Olsson and other West Point loyalists fought tooth and nail against, it was a decision that doomed the West Point mill because of the simple fact that it leveled its importance to Chesapeake with every other Company branch spread across the world. As soon as the West Point mill was equal to all the others, it lost its special advantage of being Corporate Headquarters. Worse, in time its special aura of being the birthplace of the Company no longer mattered.

Some say the move eventually came about simply because Fox's wife, Carol, did not like living in West Point. It would not be difficult to imagine that a CEO's wife might prefer to live and raise her family in Richmond instead of West Point.

"We blamed her for it," Tau Crute said, explaining how the community had loved Shirley Olsson beyond words but perhaps never really warmed up to Mrs. Fox. In all fairness, it would indeed have been difficult for anybody to have filled the shoes of the beloved "Dr. Shirley."

A closer look at truth was probably centered in the continuous bind Fox found himself in, carrying a kraft paper mill in the Chesapeake family that was growing with every passing environmentally conscious day to be a dinosaur. We live increasingly in a world where the public will tolerate in its midst only "clean companies." Paper mills, considered dirty, air polluting industries by many citizens, in spite of tremendous strides in cleaning up their act, are in today's high-tech world about as popular as a rattlesnake come to high tea.

There are many Americans today who believe all "dirty industries" ought to relocate to developing third world nations. These enlightened souls see the United States as a nation that sits high up on the ladder of respectability. America, they argue, should buy the dirty raw resources such as raw pulp or kraft paper from other nations and create the more pleasing products here, things like lovely, fragrant, toilet tissues and colorful, pretty, little boxes.

Another real problem that faced Fox was the huge fluctuation in the market. Paper is a cyclical commodity and orders for kraft paper depend wholly on orders for big appliances. When such orders fall, as they typically do in recessional times, the need for kraft paper, which protects such appliances as they move from manufacturing site to sales markets, falls in the same proportion.

As Fox spoke to me on July 3, 1996, he was exhilarated because of his recent success of turning the fiscal year into a record one of $1.234 billion in sales. This was quite a feat for any CEO of a budding global corporation. Though he was rejoicing, he still looked glum. "Prices for paper dropped this very week like stone," he said. In the corporate world one is only as good as the most recent bottom line. There is always a new quarter waiting to grab you on the sidelines. Fox was already preparing himself mentally for his next quarterly report.

"The number one challenge to any CEO is to make money and to do so in an ethical manner," Fox explained. The paper industry not only faces tremendous world competition for business, but it has the extra environmental concerns, too. So who needs the added pain and suffering of running a paper mill?

By 1997, the mill was sold to St. Laurent Paperboard Inc. of Montreal, Canada. This deed angered the people of West Point who had to endure the very next day the sudden shocking addition of the red and white Canadian maple leaf flag hoisted next to the American flag outside the mill headquarters.

Before long even Sture Olsson, who had had an office at the West Point mill since 1943, moved out and set up new quarters in a modular office building at the site of the Claiborne Courtney Seed Orchard just outside West Point.

The sale was an additional blow for Woodlands as it split the department in half, sending the Marine Department and the tugs and barges to the French Canadians along with the sawmill at Keysville. In addition, it set up Woodlands for future vulnerability with the possibility that its valuable land holdings and remaining three sawmills could someday be sold off too.

In my 1996 interview with Fox, I asked him flatly if he would ever sell the mill. He looked startled at my question, then said, "Now, why would you ask a question like that?" Since the mill's sale was announced the following spring, it is very possible that the deal was being formulated even as we spoke.

He then looked at me hard like the tough CEO that he was and his eyes narrowed. "I always have to put the interests of our stockholders first," Fox said. As he spoke, my heart sank, for I knew what those words would mean for the people of West Point.

Thus, American corporate philosophy was explained to me simply in 10 easy to understand words: "I have to put the interests of our stockholders first."

But Fox had a long run at Chesapeake and continued to expand and build it to what it is today, along with launching the Company into the international market. That no stockholder lost any money on the sale was noted as of the date this book went to press. Whether the same can be said of the town of West Point and the small businesses therein or the people who live there is quite another story.

As to whether the paper mill still had much interest in Fox's mind in 1996, he turned to the writer before exiting his interview and asked with quite a measure of passion, "Do you have any idea how much money there is to be made in marketing toilet paper to third world countries?"

He had me there. Writers never know about such things as this.

Chapter 17

The Sawmills

Mothers, don't let your sons and daughters grow up and run sawmills. That's because operating a sawmill must be about the toughest business there is in this world.

There is a scene in Margaret Mitchell's famous book, "Gone with the Wind," where Scarlet O'Hara, just after taking over her expired second husband's sawmill, puts poor convicts from the chain gang to work milling her wood. Work conditions were never worse than under Scarlet. It left me with a lifelong impression. Sawmills must be an earthly form of hell.

Maybe Chesapeake's sawmills were never run with chained convicts, but there is no doubt in my mind that eternal hiring challenges, grueling work schedules, dealing with daily market fluctuations and just plain unbearable noise levels make operating a sawmill a job that only a few people are cut out for.

One of these few men was William Phipps from Dinwiddie County, Virginia, south of Richmond, who helped build and start up several Company sawmills from 1977 to 1987. A wood dealer who bought and sold timberland, (Phipps owns over 13,000 acres of timberlands) he was a key player in launching the Keysville sawmill in the 1980s. He also helped start up sawmills in Elizabeth City, North Carolina, Princess Anne, Maryland, and West Point and Sign Pine in Virginia.

One of Phipps's most important duties, after overseeing the construction of the mills, was to hire just the right people to operate them. "The best decision I ever made as manager of the Keysville sawmill was to hire Bob Cason to run it in April of 1980," Phipps said with a smile. Cason replaced Phipps as manager of operations when Phipps finally retired from Chesapeake in 1987. Phipps, as a high compliment, called Cason "somewhat of a plugger." This must be someone who keeps on truckin,' no matter what the odds.

William Phipps with his dogs.

Phipps had the Chesapeake "do or die" and "work 'til you drop" mentality that built the Company to its present-day success, similar to Eastern Shore's Bob Owens and North Carolina's Bull Ellison. Cason remembered Phipps once telling him, "NO decision is much worse than a bad decision . . . . " There was a long pause and Phipps then spit out a wad of tobacco on the ground, took a long hard look at him and polished off his thought with, "But, Bob, just make sure you don't make too many of them!"

Another famous Phipps rule that Cason was trained under was, "You can have any Sunday off you wish as long as I tell you can have it off!" On top of this, Phipps had a habit of calling Cason in the middle of the night with some business problem. "It got old, that ringing telephone at 2 a.m.," Cason remembered with a laugh.

Besides interrupting his sleep, those night owl calls woke everyone in his family. "Finally, without me knowing it, my wife sat down and wrote Phipps a letter telling him exactly what she thought of those 2 a.m. calls!" Cason added.

The next Monday, Phipps called Cason into his office. "I received a nice letter from your wife, Bob," Phipps told him with a smile. There was another pause in the conversation. "And you can tell your wife I won't be calling you again any night after 9 p.m.!" Cason laughed again at the memory.

"I was so embarrassed, I raced right home and demanded to know from my wife what she had written to my boss behind my back! She never told me what she had told Phipps in that letter but, whatever she said, it sure worked. Phipps never called me again after 9 p.m.!"

Phipps enjoyed his ten-year stint with Chesapeake, especially hosting those famous "quail suppers" of 100 tender game birds freshly shot that he would serve everyone at his sister's restaurant in Nottoway County. Sture Olsson and the boys back at West Point would all come up for the big feast and fun. As far as Phipps was concerned, it never got any better than that.

Phipps was philosophical about working for Chesapeake. "I decided right off the bat I would do what I thought was best for the Company and, if I got support, well, that was fine and if I didn't, well, I would hit the road," Phipps remembered. "I was fortunate that when I took a stand on something, Chesapeake always supported me."

By the end of the 1980s, the sawmills handling hardwood were no longer profitable. The Company considered converting them so they could start processing southern pine, but converting a hardwood sawmill into a pine sawmill required the purchase of expensive new equipment. Here was the problem. Once hardwood was harvested on Company lands, the land was reforested in pine, which matures much faster than hardwood. Thus, over time, the hardwood reserves were reduced and replaced with pine.

After Jack King came in as Woodlands' vice president in 1989, Cason was made overall operations manager for the Lumber Division. Following the Company's new mandate to make a profit or perish, King challenged the sawmills with a tough proposition. "Figure out how to make the new sawmills profitable or they will be sold," King spelled out to Cason. Everyone at the sawmills knew their work was literally going to be cut out for them. Cason, never a man to run from a problem, threw all his energy into the challenge.

Cason made four excellent appointments: Rick Tyburski, manager of West Point sawmill; Bob Springfield, manager of Princess Anne, Maryland, sawmill; John Hurt, manager of Keysville sawmill; Terry Bullock, manager of DeJarnette Lumber Company and Willie Godfrey, overall projects manager.

Cason well remembered bringing his new team together for the very first time and firing them up with the difficult challenges ahead. "We're taking off on a new adventure," Cason told his people. "We need to each of us come up with a plan to win." They soon hit on their plan. They would convert the sawmills to process pine and start sawing Company pine logs and marketing the lumber themselves, thereby cutting out any middleman from the profits and reaping the rewards for Chesapeake.

There was an air of excitement at the sawmills with the new plan in mind. Cason came

The new chip mill at Elizabeth City, North Carolina (early 1990's).

up with a five-year conversion plan and convinced King to agree to the capital funding for the project. "We presented the scheme to Carter Fox and the Board of Directors and they liked the idea," Cason remembered.

"Princess Anne sawmill was the first conversion at a cost of $4 million for the new equipment," Cason said. "We were so excited about our plan and, in order to save some money on the project, we did most of the electrical wiring work ourselves. I even poured the cement floors myself."

Cason was still excited speaking of the return on the original investment, proving even foresters can get their kicks in making profits. "Sometimes our return on assets has been as high as 80 percent," he said proudly, "and the sawmills became the most profitable companies within Chesapeake!"

In 1985, they bought up DeJarnette Lumber Company in Milford, Virginia, near Bowling Green and Fredericksburg, a remanufacture lumber products firm. "We would resaw pallet lumber or transfer low grade wood to the new facility," Cason said. "That gave us an added dimension."

And then the big jolt came. In 1997, the pulp and paper mill at West Point was sold to St. Laurent Paperboard Inc. of Montreal, Canada. But worse for Cason, his little gem of a sawmill at Keysville went with the sale.

Cason shook his head. "It was a major blow to me and my people," he said. "Something we worked so hard at and put so much of ourselves in, well, it's hard to wake up one morning and see it all gone."

But regardless of such emotion still evident today from many employees, Chesapeake continues to follow the road to what is perceived by many as being the easy profits. The tissue and box business is in this category rather than the more fundamental and environmentally challenging paper and sawmill businesses of yesteryear.

Cason still feels proud to be a part of Chesapeake. "In the early days, when I first came to the Company, well, Chesapeake had about the biggest name and best reputation in the state of Virginia. I remember once when I had just started at Keysville, I needed an air compressor for the sawmill and I knew it would cost in the $4,000 to $5,000 range. So I called up a store in Richmond to order one and when it came time to discuss how we would pay for it they said . . . 'Hey, you work for Chesapeake. You don't need a credit check!'"

Cason sat back in his chair and thought a minute. "You know, it was true, we were a big Fortune 500 Company and no company had a better reputation in Virginia than Chesapeake. We were known for being men of our word, having good credit and really helping out in every facet within our communities, too."

Cason stared at the wall, maybe thinking about lighting up a cigarette. He had quit smoking just nine weeks before our interview, in order to mark the sale of Keysville forever in his mind. Then he sighed. "I just hope our Company never loses any of that great reputation we have enjoyed in the state of Virginia," he said.

Chapter 18

The Jack King Era

The wood business is hard, stressful work. It can affect one's health in very negative ways.

One of those ways is the threat of having a heart attack. This was a constant health threat to those in Woodlands responsible for meeting the daily demands of bringing in the wood. This was probably directly related to the constant stress of carrying responsibility for meeting Company wood quotas each and every day of the year.

It was not just area managers like Bob Owens or Bill Ellison who suffered. A heart attack even struck Paul Harper, operations manager and then group vice president of Woodlands in the mid 1980s.

Fellow North Carolina State forester graduate Jack King was ready and able to assume the newly formed position of assistant operations manager. The Company had hastily created this position for King because it wanted to make sure Harper got well. While Harper got back on his feet, King would be sure not to allow any slowdown in bringing in the wood. From his hospital bed, Harper told his fellow forester, old friend and confidant, "I don't care what else you do, Jack, just keep me from having another heart attack!"

With King's sudden elevation, the third star on the Tom Harris night sky team from the past, Dick Brake, had lost the opportunity to lead Woodlands. "But he would have been an excellent chief," Harris said upon reflection for this book. "The age thing did not quite work out for him." Also, by now, Brake had assumed command of the Delmarva Properties Inc., Chesapeake's real estate subsidiary, and was busy identifying and developing for sale Company land considered far more valuable than mere timberland. Brake, who had what was rarely found in a forester, the outgoing personality of a true salesman, was a natural with land development.

Still, the corporate world is filled with irony, fast promotions and sudden changes that nobody quite understands. Thus, Jack King now found himself immediately in charge of many old friends, including the Bull, Bill Ellison from North Carolina.

King had been well trained for the tough job he now faced. He had reported to Harper for years as an area manager and long before that had been indoctrinated by Bobby Owens over on Eastern Shore. What better experience could King possibly have had to prepare him for the responsibilities he now would face?

Owens, Ellison, and even Paul Harper had all suffered heart attacks. King had taken note of this and learned a valuable, lifesaving lesson. He knew from the start he had to face the stress and job responsibilities, along with never losing sight of his personal life with his family and his health. He also knew he had to create a job environment that could preserve his co-workers' health, but still ensure that the wood quotas for the mill would always be met.

King had always loved the Bull, but he knew that Ellison could not seem to put Company problems out of his mind. "He was a one hundred percent Company man. He would call me at midnight, never even think of the time of day, with a Company problem he was struggling with," King remembered, shaking his head.

"Once I watched him walk a continuous path around a motel as he tried to figure out some work-related problem during a time he should have been resting and having some fun. He could never let something go and just relax. He finally died of a heart attack in a motel," King said sadly.

King was determined to bring about a change. He wanted his managers to feel completely safe under his command. He let his people know he wanted them to speak out on what was on their minds in an open and

Jack King learning how to "balance the books!"

forthright way, and not to hold anything back. He wanted his staff to know that if they went out on a limb in a decision, their supervisors would support them.

"We're going to do our level best to do what we said we're going to do for you," King told his people. He hoped this open and trust-building atmosphere would improve everyone's business performance, and would also lay the foundation for a less stressful work environment and therefore better health for everyone.

King knew the wood procurement business from the bottom up, in the old Chesapeake tradition, every single detail of the work, from the 5 a.m. starts in the woods on cold mornings to the last groan and moan of machinery and men at dusk. Besides that, King knew every wood dealer, logger, and workman in the entire Chesapeake system.

Harper retired in 1988 and King stepped right into his job with perfect training for his new position. King said this was a time at the Company of the beginning of a tremendous change in wood usage at the mill.

"In the past, the mill was using mostly virgin wood for its fiber needs, wood that had been cut solely for the purpose of pulpwood and had not been pre-used," King said. "But in the 1990s that trend began to change rapidly." The mill was starting to use secondary, recycled fiber sources. At the time of our interview in 1996, King said the mill was using 1700 virgin tons and 1000 tons of secondary fiber each day. King believed secondary fiber use would eventually surpass virgin wood.

Those figures spelled coming change for Woodlands. Less need for virgin wood meant a disruption of the carefully woven traditional wood dealer and logger systems for bringing in the wood. But worse than that, it also meant less political clout for Woodlands at top management levels.

King was head man on a ship that was beginning to sink. Well, perhaps not actually sink, but the ship was certainly beginning to list. He knew it was imperative that Woodlands Division now find other sources of revenue for their valuable wood supplies rather than merely supplying pulp for the paper mill. He knew his number one responsibility to both management and employees was to find a way to raise the ship in the water.

In our interview, King spoke optimistically of new opportunities for the Woodlands Division as this century nears a close. "Frankly we have to find new, innovative ways to make money," he said. "We already know how to manage land, grow trees, and run chip mills. Now the question is how to pursue new businesses along the lines of what we do best.

"We're always listening to our people for new ideas," King said. "I keep my door open to everyone in the Company so I can be the first to hear someone's new plan for making profits. Some of our best ideas come from our own people."

By 1996, one of these ways was to expand the present business of exporting logs and chips. Continuing the development of Company land similar to the Stonehouse subdivision in James City County and the Colonial Downs racetrack and surrounding areas was another.

By the same year, Woodlands had begun to delve into international markets. In 1997, King and Jim Willis, technical manager and land services, made their first South American trip to scout prospective markets for Chesapeake Forest Products Company.

Perhaps Willis even studied a Spanish language book on his trip south. "?Como está usted?" Willis might have said as he landed in Chile with a big Chesapeake grin on his face. Woe to the Latinos if they thought they had a chance to outsmart the newly wheeling and dealing, Spanish-speaking Yankee foresters from Chesapeake.

This trip was a milestone for Woodlands. Before, the department was tops in its field but provincial in many ways. In earlier days, Chesapeake could well afford to be provincial. It had been easier then, frankly, to make profits because there was less competition. King soon found in the new global milieu that Chesapeake Woodlands had to be both good and savvy at the same time to meet new demands of world-wide competition. And if that meant his foresters had to learn to speak Spanish, then "Hablé Español!"

Another management decision King made that turned out to be unpopular, at least for forester Rae Ehlen and others in the wood treatment business, was shutting down the wood treatment business. This was tough to understand because it had been quite profitable for the Company up to that point.

Ehlen was typical of the exciting entrepreneurial spirit that had bloomed in Woodlands as the department sought increasingly to make profits independent of the mill. A Wisconsin native, Ehlen had picked up his forestry degree from Michigan State University and had come to the Company in 1970 as many others had done with experience from Westvaco and Union Camp Corporation.

In 1980, he had done a feasibility study on the wood treating industry. Soon after Chesapeake became the sole owner of the Chesapeake Bay Plywood Company in Pocomoke City, Maryland, Ehlen was instrumental in building a wood treating facility at the plywood plant. It was a simple deduction for Ehlen. "What wood we were selling for manufacturing garden and landscaping ties, we realized we could develop ourselves."

Ehlen was now president of the new Chesapeake Wood Treating Co. He was proud of the fact that within five years (a booming growth period when millions of Americans were creating sundecks and gardens in their back yards that required the use of treated wood) sales grew rapidly in his company from 0 to $100 million. It was an intoxicating time for Ehlen and he and his staff had big dreams for the future.

By 1982, Ehlen had helped in a decision to build a wood treatment facility adjacent to a recently acquired sawmill in Elizabeth City, North Carolina, the Foreman Lumber Company. Then started a period of meteoric growth. In 1983, the Company built its own wood treatment facility in Stockertown, Pennsylvania. In 1984, the Company bought another facility in Fredericksburg, Virginia. In 1986, Ehlen established a plant in Northeast, Maryland. The next year Wood Treating bought the Holly Hill plant in Holly Hill, South Carolina. By then, Ehlen's group was selling 300 million board feet of treated lumber a year, with Georgia-Pacific Corporation their largest supplier of white wood for treating and Lowe's their biggest customer.

Ehlen bloomed during these years of his rocketing business success, possibly a new experience for a forester who had rarely jumped into the hot fires of making huge profits. It was an exhilarating time for him and for those who worked with him. "We did everything ourselves, hired, trained, fired, managed, marketed, and even equipped our own plants." This rugged group even did a lot of their own building, actually pouring cement floors and wiring their own offices.

"We were producing a line of anything and everything needed to landscape a yard with decorative garden ties or all kinds of treated lumber. We produced wood for building decks, barns, docks, patios, playgrounds and even roller coasters for amusement parks," Ehlen said.

But in spite of the profits, Ehlen's efforts would eventually fold. Chesapeake management saw huge storm clouds on the horizon because of cash flow issues. In addition to this problem, Chesapeake eventually decided that wood treatment was not a "core" business.

Upper management believed there were two innate flaws in the growing wood treatment business. One, Chesapeake Wood Treating was buying too much lumber, which required huge capital outlays, and they were vulnerable, as was the plywood business, to abrupt market fluctuations. "Sometimes we found ourselves in a situation when we had too little cash and too much inventory," Jack King explained.

Even worse than this vulnerability for Chesapeake was the rising concern at the highest level of management that the Company was setting itself up for potential future liability because of perceived environmental problems with the chemicals used to treat the lumber. "It was one thing to deal with the EPA today and follow the rules exactly," King explained, "but what was acceptable today may not be acceptable 20 years from now. The Company had to consider that future risk."

Those familiar with working with the EPA and environmental liability and the potential for huge fines or other damages quickly made the decision that wood chemical treatment was a business that Chesapeake wanted to eliminate from their trade. Thus, the wood treatment business was sold.

It was a decision that hurt Ehlen and his dedicated staff who had put so much of their time and effort into the business. "We were a fragmented business with many small treatment plants, yet, at the same time, we were the largest wood treatment company in the business. The large corporation simply did not know how to incorporate the small entrepreneurial business into its fold," Ehlen believed.

Regarding environmental issues, Ehlen felt fear led way to selling the business to Universal Forest Products at a giveaway price in 1991. "We were looked at by EPA," Ehlen said, "and also OSHA. However, we were doing the best in the country in running a clean plant."

It was a heartbreaking decision, selling off the wood treatment facilities and retiring Ehlen early in 1991. It was a decision that causes some bitterness even today. But Chesapeake was beginning to see the course they wanted to navigate and was carefully taking action to see that the Company stayed on course.

A Company has to stick to its plan if it is to succeed. Woodlands, under Jack King's leadership, saw that future profits must be tied to areas of existing strengths and with as few avenues for future liability as possible.

Along the avenue of social change, King had also brought the first women foresters into Woodlands, regardless of any possible remnants of concern from Tom Harris that "hell might freeze over" if the gals ever came. In the mid 1980s, Paul Harper gave the "high sign" and Jack King hired the first permanent female foresters. It was easy enough to do. Chesapeake already had in place a strong college intern program with forestry students working each summer with the Company. With such exposure to women foresters, it was simple to see which gals would be compatible with both the Company program and the staff. The first full-time woman forester at Chesapeake was Betsy Hyde at the Warsaw, Virginia, office, followed by Amy Bigger at Keysville and Pam Leary in West Point.

King recalled that Chesapeake's hiring its first women foresters was a big event in the outside world where women were breaking gender barriers in every profession across the nation. It was also such a big moment in Company history that it actually became headline news in the local newspapers.

King planned his introduction of women foresters carefully. "I called a special meeting with all my foresters to tell their wives they would be working with 'ladies' this summer in the woods and to let me know if anyone had any problem with this," King said.

The men laughed a lot about all the adjustments they might have to make to accommodate the women like . . . dear God in heaven, what if one of them ever wanted maternity leave, which was "unthinkable" for a forester just 25 years ago.

King said he finally decided to assign the women to the close supervision of forester Fayette Wiatt. "We did this because Wiatt was the oldest forester on the staff at the time and already had six kids so we figured we ought to be able to trust him with women!"

It was a fun summer when the gals came in, giving everyone at Chesapeake a big shot in the arm. But Jack King worried over his two lady foresters like a mother hen. When he heard they had signed a lease to live in the house of some single man who might or might not be moving out, he picked up the telephone and complained to "Uncle Fayette."

"Tell the girls I don't approve of the living arrangements!" King shouted from West Point, like a father whose very own daughters were about to fly the coop.

Uncle Fayette returned the call shortly. "Jack," he said slowly over the phone, choosing his words carefully, "the girls said to tell you their living arrangements are THEIR business and for you to . . . er, I think, Jack, I think they said for you to butt out!"

Like all Woodlands chiefs, King has a lot of humor and charisma. I well remembered the first moment I learned about Jack King. Jim Vadas and I were driving toward Keysville to meet up with some wood dealers for an interview for this book.

Suddenly Vadas's car phone rang. He reached over and flicked on the speaker. "This is Jack King," the deep voice boomed and it filled up the car with some sort of presence. I thought it sounded like the voice of God and I prepared myself accordingly for a very important message.

"Go fix that fence down at Buzzard's tract, Jim," "God" commanded. Jim agreed, then said, "By the way, Jack, I have our new author with me and she will be interviewing the loggers for the book and we're on our way to Keysville."

There was a pause as God considered an author in his midst. After a word or two of welcome, the call ended. But I did not forget the voice. I knew Jack King must have been a

The Chesapeake Forest Products staff around 1985 included, from left, Jimmy Sears, Steve Whaley, Bob Owens, Jim Vadas, Sharon Miller, Walter Zingelmann, Marc St. John, Fayette Wiatt, Bill Rilee, Paul Harper, Jack King, Jim Willis, Jon Edwards and Larry Walton.

strong leader with lots of command with a voice like that.

Later, when I finally met King for an interview, I saw him get teary-eyed at the mention of forester Tom Tyler, who had recently died of cancer. Writers see everything but I knew King's feelings about Tyler were genuine. And I realized then that in spite of all the changes going on at Chesapeake there was still a lot of bonding among the Woodlands people, just like there had been from the start with all the chiefs, something that had come right down the line from the old man and the son.

I asked King, who is still heading up Woodlands, in spite of the big changes, if he had the opportunity to say something special for this book, what would be his message? He got philosophical as only a real forester, a person who spends his entire life reflecting on the well-being of trees, could do.

"You know, this society has experienced such radical changes in the last generation," he said. "We're no longer a land-based society but an urban-based society. I ran into a woman forester from Key West, Florida, a place where there are no huge tracts of trees. I was brought up in a world of trees. How can that forester appreciate the tree based values that I was brought up with?" he asked.

"Because people no longer live with trees, they think it is bad to cut trees," he continued. "It will be a growing task for us to educate the public to the understanding that trees are a renewable resource and therefore, careful planning protects the forests."

Then Jack King thought for a moment. Did I imagine that his eyes swept across his office and all the paraphernalia collected over his long years with Chesapeake, the hard hat,

147

the framed color shots of his family, a large photo of the tugboat "Sture" chugging full steam downriver and many beloved co-workers from a wide collection of retirement parties from the past?

"What I want to say is this," Jack said, looking me right in the eye with a sudden fierce look, as if fearing I might miss one word of what he said. "This Chesapeake organization was founded and developed with honesty, straightforwardness, and the use of the Golden Rule in all our dealings with others. As long as I sit in this chair, as long as we continue with the task of bringing in the wood, we will continue along those same lines."

I made a note of every word he said. Then something hit me from the back corner of my mind. As soon as I wrote those words down I knew they might have been said by the old man, the son or any one of the chiefs that ever headed up the Woodlands Division.

Chapter 19

Safety First

At Chesapeake, as with most American companies, safety was once carefully heeded and practiced by sober, careful workers and ignored by others. As the federal government grew, however, and became more concerned for safety practices in the work force, corporations began to establish official safety departments. These offices were responsible for generating safety programs within the company, their sole purpose being to teach safe practices to all employees.

Jimmy Sears of West Point was the first official Safety Department at Chesapeake Woodlands. A better man could not have been found to encourage safety practices among staff, simply because everyone loved Jimmy Sears. If Sears said it might be a good idea if workers did a certain thing, such as wear hard hats while in the woods, then most workers would do it, even if it was terribly hot that day in the woods and the last thing anyone wanted to put on his head was one of those darn hard hats.

Sears dated all the way back to 1946 with the Company and, in all those years, had done just about everything there was to do in the Woodlands Division. With a degree from the Newport News Apprentice School and years in the Navy, if there was a job Jimmy Sears could not handle, he never heard of it. Red Highland used to boast that his man Sears could do any job there was to do in Woodlands.

Jimmy Sears (left) presents membership in the "Wise Owl Club" to Sherman Wise and West Point sawmill manager Everett Tyree (circa 1990).

Sears went back far enough in Company history that he could actually remember the "wood chits" Elis Olsson paid his wood agents when times were lean and the cash flow was low. The chits were used at West Point stores and banks just like money. When the money came in, the wood chits would be bought back by Olsson one by one and cash would flow throughout the community again.

The fact that Jimmy Sears had learned to fly in the Navy was a big boon to Woodlands. He could fly anyone in the Company anywhere on either of the two Company airplanes. He recalled that one airplane was a six-seat Aero Commander and the other was a two-seat Cessna.

Not only Company people enjoyed this special transportation service. Wood dealers did, too. "I used to fly Bob Sales over in Lynchburg, Virginia, to all the horse races," Sears said with a grin. Whether or not anyone knew about this back in West Point is still a mystery.

Sears tells good stories about Company personnel with the ease of water bubbling in a boiling pot. He has a good memory that can recall all the wonderful tales of the people of the past. Sears is a breathing, walking, talking Company history.

"One of the funniest memories I have in all those years was once a group of us were in Dallas at a trade meeting," Sears began. "Some of the boys from back home must have caught the Texas spirit because when I happened to look up in the hotel restaurant that first night, Tom Tyler, Bobby Owens and Tom Harris all came through the door wearing cowboy outfits, boots and even 10-gallon hats!" Sears laughed. "And what a sight it was!"

It goes to show just how flexible those Chesapeake foresters could be. When in Texas, they sure would do as the Texans do!

Soon Sears was using his flying skills to take aerial photographs of Company tracts in Maryland, Virginia, and North Carolina. But having their own private pilot flying around in the skies above helped the Company in other ways, too. He was helpful in finding lost wood racks, the railroad flat cars used for hauling wood.

"Once the Company was desperately searching for some empty wood racks in order to bring in a big order of wood," Sears remembered. "I flew all over the state that day looking for some wood racks somewhere we could commandeer. I finally stumbled upon 50 wood racks off on a siding that nobody in the train system knew anything about!"

Sometimes bringing in the wood boiled down to the basic element of which company could get the wood racks first. The kind of on-the-spot work that included Jimmy Sears running a Chesapeake "skywatch" sometimes meant the difference in the mill staying open that week or closing down.

In 1971, when Tom Harris appointed Sears Woodland's first Safety Director, Sears' work was cut out for him from the start. His very first duty in his new position was to write the first official safety manual for the Woodlands Division.

"I had been involved with safety all along with extracurricular activities in town like voluntary rescue squad and CPR units," Sears said, "and it was natural for me to organize our Company's overall safety plan. At that time, accident rates in the woods were high and we felt that with special training on use of equipment, know-how and practice, most accidents could be prevented." Over the next 10 years, Sears would write 15 safety manuals for Woodlands, in order to meet expanding OSHA concerns.

"Our number one intent was to promote safety," Sears said. He remembered those years when he made unexpected quarterly safety inspections on everybody in Woodlands, in order to ensure that Chesapeake was in absolute compliance with OSHA.

Paul Harper presents Jimmy Sears the most revered award for safety excellence, the American Pulpwood Association's H. H. Jefferson Award, in the early 1980's.

Sears laughed at some of the resistance he had from workers who sometimes did not want to put on hard hats before they entered the woods, especially in the extreme heat of summer. Hard hats are extremely uncomfortable in the heat. Occasionally, the workers would resist. "You can lead a horse to the water but you can't make him drink," Sears said. "The same with safety. The Company can stress safety, safety, safety but that does not always mean every single worker will always be careful every single minute of the day."

Sears also recalled numerous experiences with OSHA inspectors and some of them were downright humorous. "Sometimes they sent inspectors that weren't properly trained on all the various equipment we used in the woods," Sears said with a laugh. "Once they wrote us up for not having a flame arrester on one of our tractors. [Railroad locomotives had to use flame arresters to prevent spreading sparks up and down the tracks.] So I had to tell the OSHA man diesel tractors didn't use flame arresters!"

Being written up by OSHA was a serious problem for Chesapeake, as well as every other company in the nation. The fines were stiff but they were not as bad as all the time, effort and paperwork that went into defending a charge against OSHA in the appeals process.

Sears was just the man to deal with OSHA and the sometimes unreasonable demands the organization began slapping on American business back in the 1980s. This was a time when OSHA and other government agencies were taken over by a more radical mentality, almost an anti-business attitude, that had been born in the 1960s. In those years, some businesses began to wonder, by the treatment they were receiving from OSHA, if the federal government thought the private sector was "the enemy" rather than the provider of jobs to its citizens and the basis for tax revenues to support government agencies like OSHA.

But in typical Chesapeake tradition, Sears made a point to get to know all OSHA supervisors and convince them Chesapeake was their friend and ally and not the enemy. Sears was perfect for the assignment. With his winsome personality and natural good humor, he was able to cajole even the most anti-business inspector. And Woodlands improved its safety record at the same time, which made all parties a winner.

Robert Geron (of the famous Geron family of which the father and all four sons worked for Chesapeake) continued the program of good safety practices when he became safety coordinator in 1988, after Sears retired. By the time he had taken over the position, concern over wood treatment plants weighed heavily on Woodlands management.

EPA had started the policy of holding companies responsible for future damages, even if the companies had been closed down or sold off earlier. The chemicals used to treat wood for outdoor use could possibly leave chemical residue in the soil on the site of the wood treating facilities. Such residue had to be cleaned up properly. Even though Chesapeake Forest Products eventually bowed out of chemical treatment of lumber and sold their plant locations, it did not sell the wood treatment plant site in Fredericksburg, Virginia.

During Geron's years, the Company made sure it was ready for OSHA's continued quarterly inspections. Geron saw to it the Company records were kept up to date and ready for such scrutiny. "But eventually OSHA lost its initial flagrant anti-business attitude that had so irritated Chesapeake and began to work hard in a team effort to help Woodlands comply with the growing number of safety regulations," Geron said.

"In the beginning years, they used to come in four times a year and flash their OSHA badges at us and say, 'We're OSHA and we're here to really sock it to you,'" Geron said. But in time OSHA started to see they were part of an overall team effort with business to help build safety and not to just harass businesses with one fine after another.

On the Lewis tract in Dorchester County, a knuckleboom loader piles the wood on the Green Team logging job of Richard Brown in 1994.

"Now they come in and if they see we may be weak in a certain area, they explain exactly what we need to do to bring us into compliance. They let us have a week or more to make the corrections," Geron explained. "An example might be, we put out a decibel meter on an employee to see exactly how many decibels of noise he experienced during a day," Geron explained. "This complies with regulation but maybe we didn't teach the parts of an ear to that employee. OSHA will let us hurry up and do that and come back later and see that we did it and then all is fine."

On top of OSHA regulations, Woodlands had state regulatory agency rules to comply with from NOSH, VOSH and MOSH (North Carolina, Virginia and Maryland) which, in some cases, were different from one state to the next. Geron said safety regulations could become quite complicated just by crossing state borders.

The fines could be hefty, too. "I saw a $12,000 fine come in one day and we appealed it and won," Geron said. But the government agency always has the advantage because the private sector, in the business of making profits, does not have the time state and federal agencies do to argue and defend against citations. "Sometimes it was easier and cheaper for the Company to just ante up on the fines than try to fight them, even when the Company felt the fines were unfair," Geron said.

But Geron said safety sometimes falls right back on the shoulders of the individual. "That's why hiring is so important," Geron pointed out. The fundamental practice of careful hiring in itself helps cut down accidents on the job.

Another area the Company is careful about is drug addiction. Company policy is zero toleration and this, in itself, helps prevent accidents. "We tried never to hire anyone with a drug or alcohol problem because this could hurt safe working conditions for everyone," Geron added.

Over the years Woodlands saw the introduction of BMP, the program known as Best Management Practices. "Woodlands implemented techniques to take care of the woods, and our people take special care to do this, even during the tree harvesting process. For example, we take care not to damage the forest with skidder tracks or we watch that we don't leave mud on the road when we exit the forest," Geron said.

STOP, initially a DuPont safety program, was another program that helped Chesapeake improve safety records. "Some years are great for safety, others not so great," Geron said. "Like last year [1997] was the best year ever for safety and this year [1998] was our worst year ever. [Robert Geron went to St. Laurent in the sale of the mill in the spring of 1997.] But we never stop trying to end all accidents on the job," Geron added.

Chapter 20

Accidents to Remember

No one who was afraid of a little danger ever went into the woods to chop down a tree or went out on a tug on a cold, stormy morning to pick up a load of pulpwood. Consequently, those who became a part of Woodlands were the sort of men or women who had a hearty spirit and strong constitution. The others stayed back at the office pushing papers behind a desk.

Working in the woods or on the rivers, even with safety always in mind, can be wrought with peril. The logging business, because of the cutting equipment like chain saws and chippers, is the most dangerous business in the world. Using a chain saw in the woods is very hazardous. Thus it is understandable that even with the Company using utmost care, over the years there have been a number of accidents and even some unforgettable tragedies.

No book on the history of bringing in the wood would be complete without some mention of accidents to remember. One tragedy still discussed among the employees occurred when Woodlands least expected it—at Christmas.

In the early years, it was a tradition at Woodlands to have the Company Christmas party on the deck of a barge. The best eats in town, plenty of good music and a full bar were the ingredients of annual good fun. When the troops had all assembled, the barge was towed out on the river by a tug. The good times would then begin to roll—and I do mean roll!

Over the years, a lot of fun and good memories came about from those annual holiday barge blasts. But one year back in the 1940s, Bill Gwathney, manager of the Company baseball team, the Chesapeakes, walked off the barge into the pitch black Pamunkey River and drowned. That was the end of all Company parties on the barge.

Gwathney's death was a terrible blow to the Company, the team and the community. He had been the manager of their beloved baseball team. The accident not only stopped all future barge parties, it caused the Company to give up its franchise of the team.

Chief Webster Custalow remembered witnessing a lot of near misses over the years in the woods as the big trees came down. Back in those days, the men used simple axes and crosscut saws. It was very dangerous work then and remains so in today's time with all the new high-tech equipment.

Once Eddie Bernoski saw two loggers, Wesley Johnson and his son, hit head on in two log skidders while hard at work in the woods. "Junior had run up on his daddy and was trapped inside the cab and it took another skidder to finally pull it off him and get him out," Bernoski remembered.

Poisonous snake bites were a constant concern, especially for those working in the woods and underbrush in the western part of the state or on the Eastern Shore. But most sightings were near misses. Sharon Miller said he never once heard of anyone at Chesapeake actually being bitten by a poisonous snake.

Still, near misses were almost as terrifying as the real thing and they left a lifetime memory. Once when Bernoski and Wayne Beasley were up in Prince George County tramping through a Company tract, Beasley suddenly turned to Bernoski, who did not have boots on, and said, "Hey, Eddie, you just stepped on a copperhead!"

Sure enough, when Bernoski looked down, he saw the snake stretched out in the woods between his feet like a long hose. He moved away fast, but he thought at the time that if that copperhead had been coiled, he would certainly have been struck. This would have been too bad for Bernoski because they were too far from their truck and medical offices to have received any quick medical help.

Bulldozers were especially dangerous because trees could come down on top of the driver or a limb could snap and poke a man right through the window of a cab. Carroll Dixon well remembered those early tractors that had minimal cab protection. When you hit a tree anything could happen.

Wayne Beasley with one of his favorite progeny.

"Once I hit a tree and a big snake dropped down on my head," Dixon remembered. Plenty of times when his blade hit a dead tree bees or wasps would attack with the vengeance of kamikaze planes.

Dixon said he and the rest of the men used to try to protect themselves on those early tractors by spreading mosquito nets around the cab and fastening them with clothes pins. After that snake hit him, however, Dixon tried spreading some wire over and around him in a crude attempt at building a protective cab. Nowadays the tractors come equipped with glass and steel barriers to fully protect drivers. The new equipment even comes with heaters and air conditioners.

But modern equipment does not guarantee complete safety. Some of the equipment is huge and the size alone can make it dangerous. "Once I was hurt on a skidder that had a set of wheels on it as high as my head," Dixon said.

Bernoski remembered witnessing a forester's fingers mashed while he was working in a cab in a bulldozer. And every forester, logger and wood dealer has seen his share of accidents from machetes, axes, saws or chain saws.

Victor Kaczmarski is retired now but he keeps a hard hat in his possession that he dubbed "Lucky." This particular hard hat saved his life once when he was working in the woods up in King William County. A 50-foot tree came down just a few feet away from him and lodged miraculously in another tree.

Kaczmarski remembered standing up and looking at the giant tree. He quickly grabbed his hard hat and put it on his head. Good thing, too, because, before he knew it, he heard a humongous "CRACK!" that sounded just like a rifle going off next to his ear. A 30-foot section of the tree snapped under pressure and came barreling down to pin him to the ground.

"It was a freak accident," Kaczmarski said. "It happened so fast. As soon as I heard the sound of the rifle, I was on the ground with the tree on top of me." There was no time to

155

react. "The first thought that hit me was if I had not put on that hard hat, I would have been killed instantly," Kaczmarski said.

Loading wood on truck, barge or railroad car is dangerous, too. Ed Tokarz remembered once he was almost killed helping to load wood on a wood rack in North Carolina when what he thinks must have been rotten railway ties suddenly gave way and the car rolled over. Ed spent six weeks in bed recovering from injuries.

Bill Goode, who spent 44 years working for the Company on all kinds of heavy equipment but never once suffered an injury himself, witnessed many close calls for others in the woods. An accident that stood out in his mind was an injury Sonny Gresham got when he was pinned against a truck with a bulldozer blade. "He suffered a crushed heel," Goode remembered.

Victor Kaczmarski holds the famous helmet that saved his life.

But the worst accident Goode ever saw in the woods was when a man operating a bulldozer came up on a stick that became lodged right in his throat. "We got him to the hospital fast," Goode said, "and I'm happy to report that with immediate medical care, the man survived."

Sometimes for a forester or logger just standing in the woods and looking at the trees can be dangerous, especially on windy days. Some of those big trees may look healthy, but they may be dead and hollow inside and more than ready to come down.

Forester Angela Hall from the Eastern Shore said she learned early on to stay out of the woods on days with exceptionally big gales. "One never knows when a tall tree may come crashing down exactly where you are standing!" she advised.

Chapter 21

Paying Customers

One important thing that wood dealers, producers and loggers require is to be paid quickly for bringing in the wood. Payments have been made for wood shipments to thousands of such providers by Chesapeake Woodlands since the West Point pulp mill opened in 1918.

Before payment could be made, however, the wood had to be measured. Wood was either measured right on site at the barge landing at locations like Totuskey and Pittman's Cove, or at rail or truck woodyards like Gretna, Butner and Keysville, or at the wood measurer's scale house at the pulp and paper mill at West Point.

As of 1989 wood no longer comes in to the mill by barge and wood arriving by truck is now weighed on scales located at the mill. Only wood arriving by rail is still measured "by stick" at Chesapeake. This is a process that has been used since the very beginning of the Company. One can go to West Point and still see a measurer who walks the wood racks measuring the wood with a special stick, a piece of chalk and a form to record the figures.

Once a truckload of wood enters the scales, a ticket on weight and volume is made up and sent over to a staff statistician in purchasing. The person who receives the ticket could very well be Gloria Foster Daniel, who has worked in Woodlands Accounting Department since 1972. "Accounts are paid out within eight days," Daniel said.

Use of the traditional stick measurers has provided a lot of joking and fun amongst the loggers. "That darn load of wood that we sent off that looked so full on our end would 'shake down' on the train trip over to West Point," was a typical lament. Consequently, wood dealers who used rail transportation naturally wanted the stick measuring to be done before the train started to roll. They could always get a better deal.

On the waterfront, as the wood was amassing in preparation for being loaded on the barges, the Chesapeake "stick man" might make his measurement and mark it with a chalk line. Who knew if that chalk line was ever erased either by a sudden hard rain or a workman who might have accidentally brushed up too close to the wood? But perhaps the stick man from the Company had a sharp eye for any such possible events and had a trick way of knowing when a mark had been removed or altered.

A longtime Chesapeake character

Gloria Foster Daniel enjoying a company party.

in the statistician office and also a very good sport was Ruth Martin. Now deceased, she once was the head statistician clerk for the Company.

Martin was a stickler for exact detail and had a reputation for not paying out one single penny over whatever the ticket ordered. She was not beyond questioning the ticket figures either, as many wood dealers came to know from firsthand experience.

Her reputation spread quietly and quickly, like that of a very strict teacher in school, across the full width and breadth of the wood producers. Word got out that nobody could trick that "Old Buzzard" back at Chesapeake and soon the nickname caught on all across the state.

If any of the men ever used this moniker in the West Point area, however, they would first look carefully to make sure Mrs. Martin was in no way within eyesight or earshot. Nobody wanted to get on her bad side.

But the secret nickname finally got out, as all nicknames eventually have a way of doing. This happened when a rather loud-mouthed scaleman filled out a wood dealer's ticket at some far outpost across the state, handed the ticket to the wood dealer and said, "There, now go mail that ticket to that Old Buzzard in West Point."

Surely the scaleman never imagined that the wood dealer would do exactly as he was told. However, the ticket arrived by post the following morning addressed to "Old Buzzard, Chesapeake Corporation, West Point, Virginia." The nickname was so well-known at the mill that the mail was immediately routed to the right party.

Nothing was ever publicly said about this strange event. Old Buzzard paid the ticket, just as she always did, and never said a word. Whether she ever noticed the name on the envelope or not was never known. But the story made its rounds among the Company employees and wood dealers. To this day, the Woodlands Division fondly remembers the Old Buzzard tale and the woman who watched the figures for Chesapeake with such a sharp eye.

To add to this story, over the passing years another tale emerged from this office. As she aged, Mrs. Martin kept a briar stick walking cane by her desk. If someone made her angry about something, she was known to say, "I'll beat you with my stick!" Before long, people in the Company began to call her stick the "Devil's Walking Stick," the name of a shrub found in native woods, which is known for extremely sharp thorns on its stem.

Mrs. Martin almost made good her threat once when a wood dealer came into her domain and had the audacity to challenge her payment to him. Woe to this wood dealer who dared to take on Chesapeake's Old Buzzard in person. There were some heated words as the man tried to reason with the lady. She became so angry that someone would question her figures that she picked up her Devil's Walking Stick and shook it at the fellow. He decided her figures were correct after all, and made a dash for the door.

So the tale grew. It hit the gossip lines in the wood dealers' network that if you ever questioned your payment from Chesapeake, the Old Buzzard would personally get you with her Devil's Walking Stick. There was no doubt about it, wood dealers had to take special care at Chesapeake.

Of course, this was all in fun. These stories indicate just how important Mrs. Martin was and the exactness of her figures that had to meet her standards. They emphasize the importance of those paychecks to the wood dealers. Chesapeake paychecks that emanated from this office kept thousands of families going over the last 80 years, just as the wood that was brought into the mill kept the doors of the mill open for business.

Daniel said the biggest changes in her work probably came about with the introduction of computers. Now everything is done by data processing. "It took a while to get those early computers up and going," Daniel remembered from their first introduction over 20 years ago. "When we occasionally lost records on the computer for one reason or another, well, the only words that could describe the situation is that it was bad."

Daniel said there has been a fair share of down-sizing in the last few years in her department. It seemed that the more high technology the office used, the less manpower the Company hired. She presently writes about 150 checks each week to wood dealers, logging contractors and sawmills and the records are all kept on computerized systems . . . all part of the constant change and progress in Woodlands.

Chapter 22

Hell Freezes Over!

Tom Harris had said it and he is still famous for it. "Hell will freeze over before we ever hire a female forester around here!" But it is important to note Harris was not the only forester in the country to say such a thing, nor was forestry the only profession that had such initial sentiment to the approaching feminine storm.

But the gals came anyway, all across the nation and also to Chesapeake Woodlands Division. The first lady foresters came under the joint leadership of Paul Harper, Sharon Miller and Jack King in 1985. And the truth was that the introduction of ladies to the once male-only foresters at Chesapeake signaled joyous new times at Woodlands. They were welcomed with great pride and enthusiasm.

Betsy Hyde, a graduate of Penn State who was hired by forester Steve Whaley in the Warsaw region, was the first female forester hired in a permanent position at Woodlands, although there had been some women interns hired in summer programs before Hyde. Other early women foresters were Pam Leary and Amy Bigger who were hired at the same time.

Amy Bigger came to her job interview after working for Chesapeake during the summer as an intern. It was a panel interview and after all the questions had been answered satisfactorily, Jim Willis remembered that a forester on the panel reached down deep in his pocket and pulled out a package of Redman chewing tobacco and laid it on the table. "Say, Amy," he said, "before we decide to hire you, there's just one more thing you've got to do!"

Amy stared at the chewing tobacco as if it were the head of a copperhead peering up at her from the table. "Gentlemen," she answered with a steely look, "I really want this job, but there are some things where I am just going to have to draw the line!"

The first female foresters at Chesapeake may have had just a bit of difficulty convincing Jack King they could make their own living arrangements, thank you, but they were quickly assimilated into Woodlands. Lucky for the ladies, Fayette Wiatt was the chosen one to help integrate them into the woods and not any of the other foresters. For surely it could be said

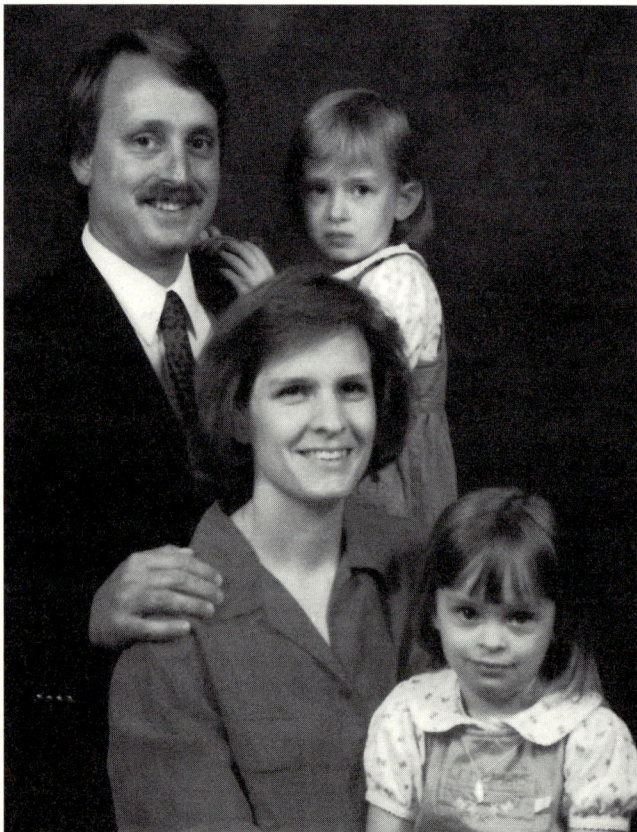

One of Chesapeake's first women foresters, Amy Bigger, proves that women can be foresters and still have a normal family. William holds Ashley and Amy is with Mary...both perhaps future foresters?

that no lady would have been safe around any of them.

Wiatt was hand selected for the assignment because he was the father of six children and the other foresters decided after a confab that after all this activity, surely Wiatt would have settled down somewhat in that department. But one never knows when it comes to foresters whether any of them ever really do "settle down" in that department. In spite of over 75 interviews made for this book, no concrete evidence on that question was ever discovered on either side of the question. There is something about trees and how they, all day and all night, continue to shoot up tiny little green shoots in the forest that might very well affect a forester in this department.

Regardless, Wiatt immediately put the new lady forester hires to work under his firm guidance with all the duties foresters do. They were soon busy in land management, burning debris to clear the land, inspecting planting and site preparation crews, and all the rest.

One of the first concerns that troubled Wiatt was the question of whether the women could hold and operate well the 12 pound burning torches that could be difficult even for the men to handle. That concern was extinguished, as well as all the rest, as soon as the ladies picked up the torches and went into the woods to work.

It turned out that women were just as good as men in forestry. Wiatt went even further than that. "Those first two gals, and all others since, have been just as good and hardworking as any men we ever had and that's an absolute fact."

Carter Fox remembered something funny that happened in Woodlands when he was CEO. Normally, management would check with the professors of various forestry schools before making a hiring decision. Chesapeake had learned through experience they could trust the faculty on hiring recommendations.

Chesapeake prided itself in always getting the "number one" graduate in forestry school. One day when Sharon Miller was in charge of Woodlands, he heard through the grapevine that North Carolina State had a "number one" forester graduate by the name of Cathey who was ready for hire.

Without a second to spare, Miller picked up the telephone and called the forestry school to tell it Chesapeake wanted to hire this Cathey fellow. Sure enough, the Cathey fellow, Joel, was interviewed and came on board. All seemed to be settled.

Then the head professor back in North Carolina State called up Miller and said, "Hey, Sharon, you got the wrong Cathey. You got Joel Cathey who was 'number two' in his class. His wife, Aurelia, was 'number one'!"

Woodlands had a big laugh. At least, with this hire Chesapeake set a new record. That was the year Chesapeake got both the "number two" and, indirectly, "number one" foresters from North Carolina State.

Today, women foresters are no longer anything unusual or special. Just about every paper company employs them, including Westvaco here in Virginia and also the State of Virginia's Department of Forestry.

Angela Jetton Hall is a present-day woman forester in Woodlands assigned to the Eastern Shore and working under region manager Larry Walton. Originally from Alabama, this gal describes herself as a "real Tomboy" who always loved to be outside and never wanted to be "trapped in an office."

Once Hall figured out she could actually be paid for working in the great outdoors, she hightailed it over to Auburn University and picked up her B.S. in forestry. She started work at Chesapeake in October of 1996.

Hall does everything a male forester does. She cruises and buys timber and manages timberland and loggers all within her team structure in the Wicomico and Worcester counties in the Salisbury, Maryland, vicinity. She is presently the only woman on that particular forestry team.

Hall never once saw herself as different from any other forester and this mentality is what it takes for women to be successful in any field previously dominated by men. "The way I look at it, if I break my leg deep in the woods, it's just as much a problem as if a male forester breaks a leg deep in the woods," she said.

What about snakes? the woman writer thought to ask (I was already quivering in my shoes). Hall paused before answering that question. Finally she responded, "The one thing that especially makes me happy about working on the Eastern Shore is . . . there are no rattlesnakes here!"

Hall is a bit on the short side and one of her first challenges with her new job with Chesapeake was learning how to stand her ground with the tough male loggers. It seemed some of the men wanted to test her to see if a woman could really do the work.

"One logger who might have been twice my size was especially resistant to taking any orders off me," Hall said with a smile. "When I gave him a tract of timber to cut one day, he up and refused to do it because the ground was too wet. I had to insist he do it. Fortunately for me, my Company team was really rooting for me and they gave me all the support I needed to stand up to the fellow. Finally he went ahead and made the cut and he did the work without even leaving any bad ruts in the soil. It was my first confrontation and I won. My team was so proud of me!"

Hall said she learned fast that no matter what, and even though loggers are independent contractors, "When they are on my land, they have to follow my orders!"

Many Chesapeake people, especially the old-timers or superachievers have found the new "team approach" to doing their work frustrating and time-consuming. But Hall likes the team approach and said teamwork is especially helpful to new employees who need to learn on the job as quickly as possible. "I can make mistakes and then ask for help from my team and its OK for me to do this," she said.

Hall talks with the bravado of a little man. She laughs and acts just like one of the boys and calls extreme environmentalists "freaks," "earth muffins" or "tree huggers." She was fortunate to know forester Tom Tyler before he died, and he taught her how to respect and also get along with environmental groups.

"We go to their meetings and hear what their concerns are and do all we can to make sure their concerns are carefully heeded," Hall said. "We learned this policy from Tyler. An example of such 'green concern' was a drive to protect the Delmarva fox squirrel, which is a big, grey, fluffy-tailed squirrel found only on parts of the Eastern Shore. In recent years, the squirrel has diminished in population and the U. S. Fish and Wildlife Service is trying to bring its numbers back.

"We now have to get a permit from soil erosion control before we can make a cut on any land," Hall said. "And believe me, it's hard telling a farmer that he can't make a cut because of the squirrel population."

But squirrels need trees for nesting and cuts have affected their ability to reproduce. Chesapeake has come up with two simple systems to help bring back the squirrels. First, the

Company has set aside special tracts of land to protect the squirrels. In addition, they have stopped all harvesting of trees during nesting times.

It is not just the squirrel that has been threatened on the Eastern Shore. The bald eagle has also experienced problems and is trying to make a comeback. Angela said waiting for their comeback is not really harmful to Chesapeake. "All land that is set aside today for the protection of wildlife will be of more value in the future when the animal population is secure once again."

Just how did adding female foresters to the Chesapeake staff really change anything? Jack King told a story that perhaps illustrates best exactly how the gals added to the mix.

"One day Angela Hall was walking through the park with her husband in Salisbury, Maryland, when a storm suddenly hit with great force. The water came down so hard and so fast the park walkways were deluged with water. Hall happened to notice a mother duck was walking with her long string of ducklings tailing behind her. Suddenly a gush of water swept the babies off and away down a raging drainpipe.

"Hall and the mother duck looked down into the drainpipe and saw the ducklings trying to keep swimming in the raging waters. She could think of no way to save them so Angela ran to a pay phone and dialed 911. Sure enough, the rescue squad showed up and saved the ducklings!"

Now, be honest. What male forester would have done that?

The St. Laurent waterfront in 1999.

Chapter 23

On the Waterfront Today

The Marine Department, a longtime and proud mainstay of Chesapeake Woodlands, was sold in 1997 to the Canadian company, St. Laurent Paperboard, Inc., along with the mill at West Point and the sawmill at Keysville. It was a sore slice that divvied up the beloved cake.

Even today just a glance at the old tugboats, the "Sture" and the "Elis," when they occasionally are docked at port at the foot of the Pamunkey bridge in West Point triggers emotion in many a soul who personally knows of their hard work and devoted service to Woodlands. Such poignant feelings will live on and on until the last tug, the last pilot, the last crew, or the very last man waiting on the shore for a line to catch from an arriving tug departs this earth. After that, this feeling will still live on and on, if only in the pages of this book.

Indeed, just seeing the old tugs at dock or a flash of green or red pushing or towing a barge back and forth through the Pamunkey River bridge or down the York headed out to the Chesapeake Bay brings back memories of so many past pilots and crews, so many loads of pulpwood and chips and so many journeys up and down the river. We won't forget the tugs, the barges, and all those loyal crews of yesteryear who worked the Woodlands water-front.

But even with a new Company and new boss, Mathews county native Tommy Callis, the marine superintendent, runs a mighty fine ship. Starting in 1971 as a simple deckhand who

164

happened to have a college degree, Tommy Callis proved early on he was a natural leader and soon took over the waterfront, following the able guidance of previous superintendents Milton Paul and Elmer Kurfmann.

Tug trips today consist of chip runs to Pocomoke City, Maryland, and Elizabeth City, North Carolina, and back to the mill and a weekly oil run to Norfolk or Newport News, Virginia, and back to fill up the fuel barge that will provide the needed fuel for the tugs for the week. They run 24 hours a day, seven days a week. The tugs rarely sit at dock.

Life on a modern day tug is grueling, hard work in all kinds of weather, with a four-man crew working six-hour shifts around the clock, 14 days on and seven days off. This schedule is exhausting and drains a man of all energy, and even a rare overnight guest aboard the tug feels the fatigue of constant, draining work.

Marine superintendent Tommy Callis on a duck hunting trip.

A tug runs wide open between jobs, nosing one barge after another into the right berth, "hot dogs" under full steam from barge to barge during its grueling work day, never knowing a leisurely moment between tasks. A tug is a worker ship, tough as nails and mean as a snake. Trying to get comfortable on a tug is like trying to cozy up to a rock.

One reason a tug works so hard is economics. It's expensive to run a tug; it costs around $150 an hour, 24 hours a day, seven days a week. The biggest problem Callis has is to coordinate the tugs with the daily departures and arrivals of trains and rush hour highway traffic going over the Pamunkey River bridge. The Marine Department presently owns two tugs and nine barges and transports a significant portion of the chips to the mill.

When asked whether the Marine Department ever experienced any emergencies, Callis threw back his head and laughed. " 'Most every day!" he answered. "A mix of boats, bridges, water, men and weather are always trouble."

But the work system, at least on paper, is simple enough. "Load up, leave port, and come back to port," Callis said. "But as easy as it sounds, somehow things seem to go wrong along the way."

According to Callis, the very first question to answer is should you pull or push the barge that day. It all depends on the weather. The Marine Department prefers to push because a tug makes better time and has better control when she pushes the barge.

However, a tug will occasionally tow a barge when the weather is rough. Sometimes the crew has to make the change from push to tow after they have gotten under way and the winds have already picked up.

Sometimes a tug will both push and pull barges, or sometimes the tug will pull two barges at the same time. It all sounds so easy, like arranging plastic ships in proper order in the bathtub, but when the wind begins to howl, the barges can break loose, get beached and even sink.

Callis had a way of describing his problems in easy terms. "Two problems can arise with barges. They can break loose and they can sink. Neither problem is easy to solve, especially in storm conditions.

"Once we had a barge break loose on us while we were pushing two barges. We had to beach the controlled barge before we could go after the runaway barge that subsequently beached itself. It took us three days and 3000 feet of hawser [rope] to pull that barge free," Callis said.

The worst problem is when the barges sink. "Once we were taking on water in bad weather off Nanticoke, Maryland," Callis said. "So we beached it and took a new barge over to shift the load. Later, when we tried to pump out the water, we found it was leaking in faster than we could pump it out."

Even a tug can sink. In 1994, the old tug "Chesapeake" sank right at the dock in front of the mill. Although it was refloated, there was some serious damage to electrical and other components. The tug was never used again. There was some talk before the sale of the mill to St. Laurent that the "Chesapeake" should be restored and turned into a museum. But funds for such an operation were never available. So the tug was sold to a private party and it is now in the process of being fully restored.

Callis mentioned that barges don't have to be in a storm to sink. They only have to be taking on water. This can happen right at the dock. To add to this problem, generally green or fresh wood chips don't float. "Once we had an 81-foot barge sink right at the dock with a full load of chips on her."

At least chips don't cause pollution problems. But an oil spill has the potential for creating a major environmental problem. Once or twice a fuel line has burst and a couple gallons of oil were lost, which is not such a serious problem. When the "Chesapeake" sank at the dock, she had oily bilge water but they managed to get her up fast enough to prevent any big problems.

"The most oil we ever spilled happened when a 55-gallon drum blew up and we had to notify the Coast Guard and EPA right away regarding that spill," Callis said. "We have a plan in hand to deal with any oil spills and we have a mock emergency spill drill every three months to make sure we are always ready to deal with any spill. We are even prepared for a 'worst case scenario' which would be the loss of a fully loaded fuel oil barge. When it comes to that threat, believe me, we are constantly training our staff to prevent such a catastrophe from ever happening here."

As in every other department of Woodlands, there have been many changes in the Marine Department over the years. Callis said probably the biggest change he has seen in recent years was the cutting back from the once traditional eight-man crew to a lean four-man crew.

Corporations, of course, exist in a capitalistic society in order to make profits because without profits they can not survive. They, therefore, do not exist to provide as many jobs as possible to the people. Thus, corporations need to keep a constant eye on the bottom line. The cutback in crew boiled down to the simple fact that running eight-man crews on a fleet of tugs became too expensive for the Company to continue. Consequently, new ideas for more cost efficiency were explored.

In earlier years each tug had a captain, a first mate who could relieve the captain on six-hour shifts, two engineers, a cook and three deckhands. Everyone knew his job and did it. Work was structured. There was little interchange of duties.

The new configuration, however, called for a captain, first mate, engineer deckhand and cook deckhand. The new smaller team calls for a new team spirit to ensure that all job responsibilities are covered. It may be cost-effective but it is exhausting work done at a horrendous pace that soon runs a man to the ground.

In the tug boat business, no one starts out at the top. All pilots and mates start out with the Company as deckhands or cooks and work their way up. This way the eventual captain has firsthand knowledge of every job on the tug.

The work is hard, the hours are long and the days are monotonously alike. "The best thing we look forward to is a good meal," said first mate Bob Mercer, who retired at the end of 1998. "And the worst thing that can happen on a tugboat is to have a lousy cook," he added with a big smile, as if he knew exactly what he was talking about from firsthand experience.

The present pilot of the "Sture," Dan Bohannon, started out as a cook and deckhand but showed so much interest in running the tugs that Mercer began to help him learn how to pilot a ship. At least that was the official line for Bohannon's promotion. But I can't help but wonder if Bohannon weren't such a lousy cook that the rest of the crew saw to it he moved out of the galley as fast as possible!

Bohannon learned to be a pilot by on-the-job training. "Mercer took me under his wing, so to speak," Bohannon said, "and showed me all the ropes. I couldn't have made it to pilot without his help.

"Not everyone would have done this for a younger man," Bohannon added. "I'll always appreciate what Mercer did for me."

Callis has tried to keep full crews on each tugboat, but as tug work wanes with less round wood and chips being used at the mill, Callis has occasionally heard those much dreaded words from his boss . . . "Cut the crews down."

Callis tries everything to keep his men at work. "Sometimes, if we have to make cuts, we try to transfer our men to the mill or some other department," Callis explained. "Sometimes we try to experiment with the work shifts, like the usual shift is two weeks on and one week off. If we can change over to one week on and two weeks off, then at least we can keep more of our men working and eligible to continue receiving benefits," Callis added. The Company has been generous with paying 100 percent of retirement contributions and also offering early retirements, which most men happily accept.

Another cost-effective change in the Marine Department was getting away from the old rule that every man had one job and no other. Now each man can do every job, from operating heavy-duty trucks, holding licenses to work on an oiler, welding, repairing small engines, to even operating the crane equipment. This has made waterfront staff more flexible and able to continue functioning even when people are out sick or on vacation.

Chesapeake tugboats help move a portion of the new Coleman Bridge into place at Gloucester Point, Virginia (circa 1995-96).

Callis said his department also learned how to cut corners in other areas. Before, whenever they needed supplies like new ropes, tools or batteries, someone would just call in an order. "Today we call for four or five prices on every object before we make a purchase," Callis said, "because it all boils down to the bottom line and if we are going to compete with the rest of the world right here in West Point, Virginia, we're going to have to control costs."

Callis uses one simple question to check his day to day decisions. "What's best for the Company?" he asks himself. Oddly enough, the answer to this question usually works out to what's best for the workers because a company that makes profits can retain jobs. "Usually what's best for everyone translates to keeping a careful eye on the bottom line," Callis added.

One area that has caused a public relations problem for the tugs and the mill is the many times the tugs or barges have hit the bridge moving in and out of the Pamunkey River. "We just paid $450,000 in repairs to the bridge," Callis said with a smile in his 1996 interview. "I guess you could say over the years the mill has kept the bridge in good repair!

"Most of the time, when a tug or barge scrapes the bridge, very little damage is done," Callis explained. "But once, in 1989, we hit the bridge twice and did we hear it from the community!" At that time, Callis rose to the occasion by announcing to those who complained, "Look, we came in and out of that bridge the last 3,782 times and never touched the bridge. Don't we get any credit for that?" Callis laughed. When it comes to tugs hitting bridges, he learned fast that an old favorite football quarterback axiom is absolutely true: You're only as good as your LAST pass!

But occasionally a "man bites dog" story occurs, as when a bridge actually hits a tug. "This happened to us once when an 85-foot span of a hydraulic bridge dropped on one of our tugs down in Elizabeth City, almost killing our pilot," Callis said. "Now that made the news."

Another favorite tug story at Chesapeake happened in the mid-90s when the "Elis" was coming back from a trip to Norfolk on her last fuel run before Christmas. As she was coming through the Jordan Bridge on the southern branch of the Elizabeth River, she called the attendant to open the bridge.

Suddenly a man in a white van who had perhaps been celebrating the holiday "spirit" in more ways than one, came crashing through the wooden gate barriers. To the astonishment of the witnesses, he fell off the bridge and into the water, about a 15-foot fall.

Bobby Williams was pilot of the "Elis" that evening, Rudy Katshinar was first mate, and Tom Scott and Steve Healy were the engineers. When the man surfaced, Williams gave the signal to start the "man overboard" drill. Katshinar jumped into the water and rescued the man from drowning.

The U. S. Coast Guard later presented Chesapeake, the "Elis" and her crew and especially Rudy Katshinar with its highest life-saving award. "Saving the life of another person at sea is certainly one of the Marine Department's proudest moments," Callis said.

Some stories about life on a tugboat are not so serious but rather full of fun. A favorite story that fits into this category is from the crew on the tugboat "Sture" and is told on Captain Bo Traywick of Deltaville, a VMI grad with a proper military background, who is a stickler for discipline at sea.

As can be imagined, Traywick ran a mighty tight ship. But the crew reported that one day the Captain went over the line when he decided the "Sture" crew should appear for work dressed properly in uniforms suitable for a tug crew. Traywick showed up the next day in military khaki dress including epaulets on the shoulder.

The men stood around in their jeans and T-shirts wondering what to do with the new brand of "spit and polish" on the "Sture." They did not know whether the new uniform called for a salute or what. After a quick confab, they decided to just ignore the whole thing and continue to do their work as usual.

"We didn't know what to think of our skipper," said one of the crew sheepishly. "All we knew is we sure as heck didn't want to wear any uniforms ourselves. We figured if we all kept quiet and went about our work dressed as usual, in time the uniform phase would pass."

It did. After a month of no results, Traywick finally gave up trying to upgrade the dress on the tugboat "Sture." He returned to regular working tug dress and was greatly appreciated for it.

It is perhaps part of the future that the great days of tugboats are beginning to recede. Trains and trucks are waiting in the wings to take over cargoes. Even as I type these words, 70 percent of all finished products at the mill leave West Point by truck.

But the waterfront is as

The "Sture" heads out to sea on a new journey in 1999.

exciting as ever. Whether it be a returning tug slipping through the bridge at dawn with a line of stopped traffic on each side waiting to pass, their headlights spread out like diamonds in a necklace, or a tug breaking through ice on the way out to the Coleman Bridge, its steel bow sounding like the scratch of a cat's claws on a blackboard, the waterfront is exciting.

And plenty of Chesapeake people who once worked on the tugs are still around to prove it.

Chapter 24

New Trends and a Bright Future

This is a book that has addressed change in one small corner of the corporate world in America. The rule most evident from all we have read is this: In today's times, constant, ongoing and rapid change is taking place, and even to be expected, in every layer of business. The rule is severe. Competition demands allegiance to the rule. Companies that won't or don't change, don't survive.

Chesapeake Woodlands has been part of that change. As the years have passed, no department or individual has been left to do work the way things were done in the past. It did not happen and it never will. For change is a part of every facet of life.

Change even affected the way people were hired. Gone are the days when Chesapeake Woodlands merely hired a son or daughter of a faithful past employer from what was commonly known in the West Point area as a tried and true "Chesapeake family." All new hires today are made through a modern Human Resource Department that considers all applicants on an equal footing. It is not unusual today to see adult children of two-and three-generation Chesapeake families seeking employment elsewhere. This was something unheard of just 20 years ago.

Also gone is the Chesapeake practice of hiring just the right person in each territory, a strong leader to crack the whip or do whatever else it took to complete the task of bringing in the wood. Woodlands now uses the Chesapeake "team approach," as do many American corporations that have found that the only way to compete in a global economy is to include every employee in the inner circle of responsibility for making a profit.

Technical and land services manager Jim Willis said the team approach has its strengths and weaknesses. But a definite strength is that it allows people at each level of work to make their own decisions as to how to solve their own specific problems. The idea is that no one knows better how to solve problems than those closest to the problems.

In yesteryear, most major Company problems were solved at the top. Decisions were meted out to all others who more or less accepted these decisions or packed up and went off to look for another job. But today's more enlightened view is that there is no one better suited to identify and then solve problems than the people directly concerned with that problem. In other words, if loggers are leaving ruts in the woods with their equipment, who but the loggers themselves should be involved in figuring out how to eliminate those ruts?

Diversity improves the team. The more variety of personalities included in a team, the better that team will function. According to Willis, foresters tend to be a bit introverted by nature. If too many introverts make up a team without a balance of extroverts, a team can be somewhat hampered.

A possible weakness in the team approach is its tendency to slow down decision making to sometimes almost a snail's pace. At one time, a decision could be made quickly by Tom Harris, Sharon Miller or Paul Harper. But today, problems may have to wait for the team to address them. Occasionally, the team can talk the problem to death before any real action is taken.

Tommy Callis, marine superintendent, explained why the team approach is not always effective in his line of work. "We have a barge sinking. Can we meet in a team and discuss what to do with the sinking barge? No. The barge will sink as we talk. Then we will have to meet to discuss how to pick it up from the ocean floor!"

Another weakness of team decisions is the possibility that a team does not want to take a big risk and will actually act as a group to cover itself from any responsibility or liability. Sometimes a team will unconsciously shut down highly productive, autonomous, creative, over-achievers who become highly frustrated having to work within a slow moving team.

Worse, sometimes a team generates "game-playing," pretending that the team is solving problems when in fact the problems are being solved by the leaders just the same as before. Only now the leaders must manipulate their teams in order to take proper action to get desired results.

But the strengths of teamwork are many. One, it helps new employees get on board quickly and obtain real on-the-job training. It also helps them assimilate into the "old boys' network" and become part of the group. Then, it offers many heads of wisdom and experience instead of just a few. With problem solving, many heads are better than one.

The team approach also delivers full support from the group when a decision or policy is made. Teamwork pulls everyone working for the Company into direct responsibility for running the Company and making a profit. Finally, by making all employees responsible it stops "blue-collar" versus "white-collar" thinking and any possible leftover attitudes from yesteryear that something "is not my job" or "not my problem."

Computers have also changed everything. "You can't even get a new screw if you don't put it on the computer first," Callis said. All land records are computerized, a far cry from the days when Tau Crute used to know every inch of Chesapeake land right in his own head and the plats might or might not have been easy to find in his office.

Today's business environment is not just the day of government regulations, but the day of constantly changing government regulations. Woodlands now has to hire people for the sole purpose of keeping up with the rapid changes of regulations.

"Paperwork is a constant problem," said Callis, shaking his head. "These days, it seems like you have to fill out a form before you make any move."

Public relations are as important as ever, even though employees are spread thin in that department. Every Chesapeake employee has to take on the responsibility of PR in today's world.

"We now have to take the time and patiently explain everything we do to the public," said Jim Willis. At one time Woodlands hired two full-time people to do Company tours, visit the schools regularly, and even take children on a tugboat.

"You should see how much the public loves our tugs," Callis said with obvious pride. "Every year, during the West Point Crab Carnival, we are overrun with children who want to get on the 'Sture' or the 'Elis.' And every so often a busload of college students from the College of William and Mary or Rappahannock Community College arrives on the waterfront just to see the tugs."

But the most important changes, according to Carter Fox, have to come from the next CEOs. "What I brought to the Company was strategic thinking along the lines of financial skills that put Chesapeake in a good financial position for economic growth. The next leader needs to get us into the world market and fast," Fox said.

"Wisconsin Tissue woke us up to the new global markets," Fox added. "Getting into Mexico and France will continue to wake us up. We must become a global company that serves global customers. The next CEO needs to be more of a global marketer and go for a world-wide position before it's too late to get into the game."

Lastly, this book has attempted to highlight some of the many achievements of the Woodlands Division of Chesapeake Corporation since its beginning and right down to the present day. Having started this book project in 1996, we now bring to a close this last chapter as of July 1999. We have discussed all the changes that have come about through the years in Woodlands. But in this three-year time period alone when this book was written, many more changes have come about in Woodlands.

Change does not frighten us or leave us cowering in fear at its front door. This is Chesapeake country. Its people have always been as strong, tough and resilient as its beloved founder, Elis Olsson, so many years ago.

We are excited about change. We stand tall and straight like the trees that we know so well in the forest and we look toward the future with what only can be called a great sense of excitement.

The future, like our past, is truly exciting. Chesapeake Corporation is moving quickly to gain and secure world markets in recycled tissue and packaging. The Company now has over 40 locations around the world in the United States, France, Mexico and Canada with plans for even more expansion.

In January of this year, a plan to build a new tissue mill and converting facility was announced by president and chief executive officer Thomas H. Johnson. The site chosen is on the south bank of the Roanoke River in Halifax County, North Carolina, the hometown of Tom Harris and Dick Brake. This new facility will position the Company near large supplies of recyclable waste paper, the basic raw material for Wisconsin Tissue Products.

The project will cost the Company $180 million and will be up and running by 2001. Now that is exciting. A new paper mill. A new product to market.

Another exciting expansion is the Company's bid to buy the Field Group in the United Kingdom, which is a company that designs and produces cartons, containers, leaflets and labels. The Field Group has 17 facilities in the United Kingdom, Ireland, the Netherlands, Belgium and France. This new acquisition will help Chesapeake develop a pan-European supply network system.

Because the assets of Chesapeake Woodlands were sold this spring to Hancock Timber Resources Group, the Woodlands Division will no longer exist as this book goes to press. But we celebrate this great family, from Winslow Gooch all the way to Jack King. And we remember these dedicated people and the Company with which they once were associated.

Who knows what the next chapters will contain for the hardworking people of Woodlands? Only the passage of time will tell. But I have a feeling, knowing these special people as well as I do, that the most exciting chapters in the book are yet to be written.

Godspeed.

Amen.

"The Big Move" cartoon was created by an unidentified artist when Chesapeake's corporate offices moved to Richmond.

Appendix I

A memo to the author from Chesapeake Forest Products Company

With the sale of the West Point pulp mill to St. Laurent Fiberboard in May of 1997, Chesapeake Forest Products and its employees entered a period of uncertainty and high stress levels. The old tightly knit Woodlands organization was split down the middle with approximately two-thirds of the employees going to the mill buyer, St. Laurent Fiberboard. No one had an opportunity to negotiate or appeal an assignment, and everyone wondered where the new Chesapeake Forest Products was going. Within just two years we would find out.

One of the first things Woodland employees had to learn was the marketing of wood. Provisions for the mill sale included a fiber supply agreement which provided that Chesapeake would supply St. Laurent with wood and chips each year. Immediately, disagreements arose over pricing methods, wood merchandising and harvest quality. None were serious but considerable negotiation was needed to settle these issues.

Another early need was office space. Chesapeake Forest Products Company (CFPC) employees were supposed to be out of the old Woodlands offices on 15th Street in downtown West Point, Virginia, by September 1, 1997. Construction superintendent Jim Vadas began what only could be called a mad scramble. Offices were added to the old Forestry garage near the Claiborne Courtney Seed Orchard and a modular trailer was hauled in. (Forest Products soon became known as the "Double-wide Department") Unfortunately, construction work never proceeds on schedule. Consequently, a number of employees operated out of cardboard boxes for a month during the fall of 1997. Many other boxes were never even unpacked before the land and sawmills were sold less than two years later.

As financial and operating plans were prepared for the new Chesapeake Forest Products Company, it soon became obvious that former successes as a pulpwood producing company would provide problems for the new "timber" company. During the preceding decade, every unusually wet winter caused many timber stands to be harvested just before they began to produce significant amounts of high value sawtimber. This may have kept the mill operating at the time, but it greatly reduced the amount of sawtimber that could be produced in the near term. It was quickly decided that the only way to keep up income was to sell land. Many of the younger foresters wondered aloud if they would outlast our rapidly diminishing land.

Any doubts about the fate of Forest Products quickly evaporated in the late summer of 1997. Carter Fox, who always had a keen interest in forestry, abruptly retired as president of Chesapeake Corporation. He was replaced by Thomas Johnson. Johnson had most recently been with Riverwood Corporation and had just completed the sale of their woodlands to Plum Creek Timber Company. His focus was packaging and tissue and he needed money to grow those businesses.

In late 1998, Shaw, McLoed, Belser and Hurlbutt, a consulting forestry firm from Sumter, South Carolina, was hired to evaluate all of the land and timber owned by Chesapeake. Shortly thereafter, a few forest products companies began to visit Chesapeake's operations. In May of 1998, Bill Tolley, Chesapeake's senior vice president, told employees that a decision on CFPC's fate would be forthcoming in "90 days." The decision had already been made to sell, of course, but it took over a year to find a buyer.

In late spring of 1998, an investment firm that quickly came to be called "Golden Socks" was hired to shepherd the sale. Its first draft prospectus was a premonition of what was to come. The report was replete with "West Coast" forestry terms, factual errors, and general misunderstanding of forestry. Forest Products employees who were assigned to help with the sale made a major rewrite of the document. Later, as the financial wizards began bringing prospective buyers to the actual operations, it was necessary to explain to them such terms as "logging systems," "site preparation" and even "chiggers"!

Another error in merchandising the land may have been made during the early summer of 1998. Company foresters were asked to prepare a list of tracts with potential real estate value. Thinking that we would be the ones helping in managing and selling the land, we prepared a list of land that "might" be sold in five to 10 years. The list, which contained some 40,000 acres, was abruptly pulled out of the sale package and identified as "Higher and Better Use" or "HBU" lands. This removed some older timber and some of the best timberland from the sale package. Several buyers had seen some of the land and realized that their near-term income would be reduced.

Eventually, five organizations became serious suitors for Chesapeake Forest Products. Jim Willis, operations manager, developed a tour route and booklet that could be tailored to the desires or needs of whatever company was visiting that day. Usually it was a two-day visit. The sawmill specialists would visit all three of our sawmills. Then the woodlands experts would visit the selected tracts in the West Point, Keysville and Pocomoke City regions. Almost two full days were spent in such field visits. Then, an indoor presentation of the visits would be made to Corporate officers in Richmond.

After the second or third trip, these tours attained a certain predictability. Chris Burgess, assistant vice president, was the official van driver. Chris was going through a "mid-life crisis" and had just bought a very fast sports car. On the back roads, he often drove the van just like his new Stealth sports car. Each time someone's head had to be scraped off the ceiling after hitting a bump at excessive speed, Jack King would relate a story of suddenly seeing some distinguished vice president with his hat jammed down over his ears.

For some reason, West Coast visitors thought the only way to see the land was by helicopter. (Maybe they had been warned by East Coast visitors to stay out of all Company vans!) Each time they took a "viewing," they saw little but treetops, a wide selection of Tidewater birds and the Chesapeake Bay. However, as nice as such trips certainly are, to appreciate the true quality of timber, it must be seen from eye level and not 200 feet above the treetops. With no ground transportation available at each stop, (heliports) the timberland could only be examined in detail for 200 yards around the heliport. However, the arrival of a helicopter and its passing overhead (at 200 feet!) always stirred considerable animated discussion among CFPC employees, who were way past the "90 day decision date" promised by Bill Tolley.

Usually, the Woodland visits accomplished little except to make "Golden Socks" feel useful . . . and richer. The more perceptive buyers knew that they were being shown the best land and timber; the less perceptive did not appear to have serious money sources.

The summer of 1998 became particularly stressful for CFPC employees. David Birdsall's first question each morning upon arrival at the office was, "Well, who's looking at us today?" Later, after the visits were finished in September, the question became, "When will we know something?" The standard answer was, "Probably within a month." That "one month mantra" was repeated for seven more months. Finally, on April 15, 1999, it was announced that the sawmills were being sold to St. Laurent Forest Products Company, while the land was being monetized to Hancock Timber Resources, an investment arm of John Hancock Life Insurance.

Like all epics, this is a story of a people and their land. As of September 15, 1999 the people and their fates are as follows: Of 180 employees at the time of the sale 163 Lumber Products employees were offered equivalent positions by St. Laurent, 1 forester became an employee of Resource Management Services, no foresters became Hancock's forestry consultants, 8 employees took elected early retirement and several employees started their own consulting business.

Time will tell the fate of the land. Hancock has a good record of land stewardship but is at the mercy of its investors. The 55,000 acres of HBU land was retained by Delmarva Properties, the real estate arm of Chesapeake Corporation. All of it, including two of Chesapeake's very popular nature trails, should be sold by 2005. All of it is scheduled for small parcel sales, but Chesapeake's foresters can only hope that the new owners will maintain at least some of their land in timber. After all, all foresters know there is no higher and better land use than growing trees!

CFPC

September 15, 1999

Appendix II

Chesapeake Has Real Live World War II Hero

by Mary Wakefield Buxton

URBANNA, VA—Every American corporation has a real live hero amongst its ranks and our Chesapeake Corporation over in West Point, Virginia, is no exception. The reason I know about Chesapeake's hero is I am writing a book for the Woodlands Division and in the process interviewing many of its past and present people. I am learning how pine seedlings grow into big trees and eventually become paper and wood products and thus have borne an entire civilization on their limbs.

But better than that, I am meeting some of the loyal people who were or are the mind, heart and soul of this Company. There is nothing in this world more interesting than people, and the people of Chesapeake are no exception.

Last week while in Lynchburg, Virginia, I met Bob Sales, a retired independent wood dealer who supplied Chesapeake with wood for over 40 years. He was a key player for the Company from central Virginia. He also was the only survivor from his company that landed with the U. S. Army in Normandy in 1944.

I was so fascinated by his real life war stories, I could hardly hold onto my pen. "Tell me everything, Bob, don't hold anything back," I said, trying to look like I wasn't about to fall off the edge of my seat. He did, and in such vivid description, I felt like I was on that landing beach, too.

Halfway through his story, when he told me how he managed to make it to the beach and saw his wounded radio sergeant a couple hundred feet away, and when this friend had raised his arm to wave for help, a German sniper picked him off from the cliffs, his head seeming to explode on that distant stretch of foreign sand and, well, . . . I burst into tears. Trouble with this writer is I felt the bullet, too.

Oh boy, did Bob Sales know he had a "live one" in me. Try as I might to appear to be a sophisticated journalist, it just didn't wash. I sniffled in my hanky like a child. This was, of course, only a shrewd feminine ploy. I was just trying to get him to open up and give me more of the real scoop.

Sales was on one of the first landing craft into shore. They had to jump into the heavy seas in columns two by two, with the nose of the boat wide open to machine gun fire from the cliffs. The depth of the sea was up to their necks each laden down with a uniform and equipment. Many of our troops who escaped the mowing-down gunfire drowned under the massive weight.

His commanding officer was first man off and was killed right as he hit the water. Two more followed and met the same instant death. Sales was the fourth man in the water and because he had somehow jumped off the side he had managed to escape the line of fire. Someone in the water helped him get loose the radio pack around his shoulders so he wouldn't sink. He looked back and saw every man coming off his landing craft hit by a stream of bullets.

Somehow Sales made it to the beach where he lay sprawled in the sand. Slowly, traveling in inches like a snail in the sand, he managed to creep up the beach for cover. One wrong move, one tiny flicker of a hand, could catch the eye of a sniper and he would train his guns on him.

Later, on the march into Germany, a lone sniper killed three men in his group in one morning. Sales decided he would get this German if it was the last thing he did. He circled around and came up behind the lone paratrooper left behind to slow the advance of the Americans. Sales pumped six bullets into the soldier.

The dying German signaled for a cigarette and Sales lit one for him and held it to his lips for one last smoke before he died. Sales found a picture of the soldier which he saved and brought home after the war. Sales placed the picture in front of me and I stared down into the eyes of a young man dressed up in uniform. He looked like my son. Would somebody please pass another tissue?

While attending the 50-year anniversary in 1994 of the Normandy Invasion, Bob Sales stands by Captain Ware's grave in France.

Sales was on a tank which took a direct hit and he was plummeted through the air with his eyes on fire and filled with shrapnel. He spent over a year recovering in U. S. hospitals and eventually was fortunate to regain his sight.

It was men like Sales who took back Europe from the Germans. They were tough men but too young to know they were caught up in a battle of the century. Only years later would Sales finally come to understand just what he and his peers had done for the rest of the world.

Sales enlisted at 16 and, like a lot of patriotic Americans, lied about his age to get in service. He had no idea what he was getting in for and was excited about going off to war. One day the company in Virginia was sent up to New York and loaded on a big ship. None of the men knew it at the time but all 15,000 troops were on board the Queen Mary.

It was 1942 and England was desperate for American manpower. Her lands had been emptied of all men between 18 and 65. They were down in north Africa fighting Rommel, trying to break through German lines from the south. If Americans had not come in to England to defend the tight little island, Hitler might have walked to Buckingham Palace with nary a shot.

On his way to England and a day out of Scotland, Sales was standing on the deck of the Queen Mary when he saw a minesweeper nearby. The queen was zig-zagging along at a top speed with her precious cargo of troops, in order to avoid the U-boats. But the minesweeper got too close. Sales looked up to see the queen slice through the other ship. He watched as the two pieces of ship and her sailors sank into the sea. The queen never stopped, never even slowed, but gave one mighty whistle and kept on steaming for England.

Sales was still young at the time but as he stared at the poor men struggling their last minutes of life in the rolling seas, he began to understand one thing my generation would learn in Vietnam. War is hell.

Still, Sales's story is but one tale of many. This is one writer's poor stab at trying to say something to this older generation, this World War II generation, this generation with so much character and courage, which saved England, took back France and most of Europe.

Thank you. And thank you Bob Sales for giving me a vial of sand from Omaha Beach. Every time I hold it in my hand I will think of only one thing—good ol' American courage.

(Reprinted with permission from the Southside Sentinel, Urbanna, VA., 1996)

Appendix III

A Visit with the King

by Mary Wakefield Buxton

URBANNA, VA—Journalists would give anything to interview a king. We wait on the sidelines hoping to be called. Kings, however, are few and infrequently found. But we do still have them. And I was called to meet the King of West Point, if not all Virginia.

His name is Sture Olsson, son of the first king, Elis Olsson, founder of Chesapeake Corporation, giver of many jobs and great wealth to Virginia. Of all the newspaper people out there, I was the one that got the interview with the king. Tough, hard-nosed, sophisticated me from the Southside Sentinel. Er . . . that's over in Urbanna, Mr. Olsson, in case you didn't know, the heartbeat of Middlesex County.

The reason I was called to see the king is I am writing a book about the Woodlands Division of the Chesapeake Corporation. A writer cannot capture the heart and soul of the kingdom without first meeting the king. I had to get up close to him, see what he looks like, talk to him, find out what he thinks and feels, how he behaves and what he believes passionately way down deep. To do this properly, I had to meet the king.

I had heard about him for years. He had started out at the mill as a young engineer in 1943 just after graduating from UVa. By 1953, he was promoted to CEO after his father, Elis, retired. By 1967 he became Chairman of the Board in a position he held with Chesapeake until 1994. He has a son, another Elis, working in the Company. Since the Company started making paper in 1917, there has always been an Olsson on hand at West Point.

Sture Olsson and the Olsson family are the stuff that myths are made of. All Chesapeake people adore them. They are tough, honest, no-nonsense folk, (the old man was originally from Sweden) who came to this country via Canada with the knowledge of how to make paper. They are the kind who have always worked hard and always will, no matter how much money they make. It's in their genes.

"If you are afraid of work, don't get into the wood and paper business," Olsson told me while his eyes drilled me like a couple of steel knives as if suspicious as to whether I might be the sort who was afraid of hard work. Me? Why, I love work. Work is my middle name.

"You can hear him coming a half mile away, just like his old man," one of his retired employees told me. "In those days you saw an Olsson every day you went to work at the mill. Those two ran this Company by walking around the plant every day and talking to everyone. They knew everyone and everything. That was how a man ran a business in those days. He got the best paper products and he got top work from his people."

But what to wear to see the king? I looked over my dull dresses and finally chose the number that makes me look like a Sunday school teacher. Sweetness and innocence would be the best ploy, I decided, to get the real scoop from a king.

On the day of the interview, I hopped in the car and sped over to his palace on the Pamunkey River. As I approached West Point I saw the paper mill emerge on the skyline. It filled the sky like Windsor Castle. It had everything—turrets, moats, fanfare and steam.

I looked in my rearview mirror and saw one of the king's men on my tail. A big tractor-trailer, loaded with full-length pines freshly plucked from the forest, was coming up fast and racing hard for the mill. I floored the speedy Toyota. So much for trying to beat the king's men. He passed me like a greyhound after a hare.

Olsson was my 16th interview out of 75 at Chesapeake Corporation. Until I saw the king up front and in person, I knew I would not really understand his people. I had heard so much about him. "He swears every other word," someone warned me as if hoping I wouldn't faint at any "colorful" words. Ha! I was an ex-navy wife and besides that a one-time mother of two teenagers. No colorful words surprise me.

"He says exactly what he thinks." Ha again. I do, too. We ought to have an interesting little talk. "No one pulls the wool over the eyes of an Olsson," another said. Ha to that, too. He hasn't seen anything yet. Wait 'til he runs into a really, tough, hard-nosed, journalist lady from Urbanna. Why, I would ask him questions he had never even thought of yet.

I pulled into the guest parking lot at the exact appointed time. The outer lobby was empty and locked, the latest corporate trick to safeguard its people from the rest of us. Two huge portraits of Elis and Sture Olsson loomed on the wall over an altar of a carpeted version of the famous Company logo, the rolling "C." A copy of Forbes magazine lay on the desk. I picked up the telephone and dialed a number. An assistant, Thelma Downey, greeted me and took me upstairs to the throne room.

She escorted me into the inner chambers. A handsome, white-haired gentleman stood up dressed in red shirt, khaki pants, and bright red socks. Gosh. The king wears bright red socks. I made a note of it. I knew I had been upstaged. Even my Sunday school dress could not compete with his socks. I had to fight back a sudden urge to giggle.

I was saved. From under the desk popped out a rust-colored spaniel. The dog headed right for me. "Who the hell are you and what the hell are you up to?" boomed the king with a stern Scandinavian look, hardly softened by a generation in sweet-talking, genteel Virginia. I smiled and threw my arms around him and kissed the top of his adorable head.

The dog, that is, not the king.

"I love dogs," I said peering up at the king with a sweet look. "I just love your baby." Such are the ways of tough, hard-nosed journalists. We do what we have to do to survive. There is more than one way to skin a king. Kissing the dog is just my skin-the-king plan number 336. It worked, too. The king melted. I told him all about my baby back home, "Tucker," the golden retriever, and we were off on a smooth track.

We talked for two and a half hours. I asked the questions, some tough enough to get the king to squirm just a bit (after all, this is the press), and I always got an honest, forthright response. The king speaks right from the heart. There is nothing disingenuous in him. I like that trait in kings. Listening to his words was a woman who sat spellbound and tried hard not to show it.

You see, I just love these old American business magnates who have given so much to our country. I love to hear them tell how the Company got started, how the business was run and how they kept the doors open even in the lean times. The Olssons knew how to do this. These kinds of people built this nation and made us what we are. They are the entrepreneurs who take the big risks and who make the world turn. They create the jobs, the money, the taxes, and the life style for the rest of us. Without these people who built the corporations, we would all be out of work or working for the government.

A wood dealer from Lynchburg told me, "The Olssons, in spite of their great success, never took on airs. They were just like us. In the early years the Company used to have big annual parties for their wood dealers. We never missed a Chesapeake party. One time I took a friend with me and I introduced him to Sture Olsson, not mentioning he was the CEO of the Company. My friend asked the man what he did for a living. Sture said, 'I'm in the wood business too.'"

Chesapeake is an unusual business. It can not stay open even one day without getting in the wood. Every worker at the mill knows that without the daily incoming wood there would be no mill and no jobs and no big money in West Point.

The story of the wood and paper business is fascinating. I am examining the Company just the way I saw one of their foresters examine a tree one day. He peeled back the bark, sniffed the moist cool flesh inside, snapped off a twig, pinched the leaf, held the blossom to the sun and watched how the light falls on the bloom. Writers are just like foresters. They examine trees. We examine people.

I want to know everything. I want to know where Chesapeake came from, where it is now, and where it is heading in the future. I want to know its people like I know every character of a favorite book.

When I have finished my book on Chesapeake Corporation, it will contain the many separate tales of the people who brought in the wood each day and also a much bigger picture. It will be the story of a part of America. Its people will be just like the rest of us, filled with the same hard work, pride, joy, hopes, dreams, fears and furies.

I kissed him good-bye.

The dog, that is, not the king.

The dog, with its waggly tail, trailed after me out the door and down the hall. The king followed. "I want to read every damn word you write and before it's printed, too!" he said gruffly. "Yes, Mr. Olsson," I answered sweetly. This king was a clever king. He knew the nature of writers. In the old days, he could have had my head. But in today's world, I could have his.

Still, I meant to humanize the king and his kingdom. I meant to tell the whole world that the king wears red socks and keeps a dog under his desk. And the king is gruff on the outside but as sweet as sugar underneath.

"Don't forget our deal," was the king's last command to me. I laughed and waved good-bye. What deal?

God save the king.

(Reprinted with permission from the Southside Sentinel, Urbanna, Va., 1996)

Appendix IV

"Sugar" Meets the Chief

by Mary Wakefield Buxton

URBANNA, VA—My work for Chesapeake Corporation in interviewing past Woodlands employees had brought me to the home of a real Indian chief. I had driven into the heartland of the Mattaponi tribe along the river and had rung the doorbell of Chief Webster "Little Eagle" Custalow, who had spent 45 years supplying pulpwood to the West Point mill.

The 84-year-old chief answered the door with his cocker spaniel at his side and bid me greeting. "Why, hello, Sugar," he said with a warm smile. I shook his hand. "Welcome to the Mattaponi Indian Reservation."

As I sat down, the dog staying at my side to receive long and lovely pets about the ear. While I conducted my interview, I could not help but wonder if a woman could become an Indian chief these days.

"Can a woman become an Indian chief?" I asked the chief, never the sort to run from a good question.

"Why, sure she can, Sugar," Custalow answered. "Why everything has changed these days. Our tribe still has a chief that passes down from father to son, but some tribes elect their chiefs. In those tribes anybody can run for chief."

My journeys in life had taken me to some interesting places, I thought as I considered the chief before me. He looked like a very wise old man who had seen a lot of life. I had never met an Indian chief before and it was a real thrill to hear his story.

"You're not a Virginian, are you, Sugar?" asked the old chief. There was no fooling him. OHIO must have been stamped on my forehead like Rose Bowl script from the OSU marching band. But I had been in Virginia since 1959. How in the world had the old Chief known I was not a native?

As I mulled this question in my mind, I looked at the framed pictures on the living room walls. Chief Custalow stood smiling with Virginia governors past and present. Here was a man who was as honored in Virginia as a prince.

"You know, the Mattaponi tribe is the oldest tribe in America," he said. "My daddy told me that George Washington once said without the help of the Indians, the early Colonists would never have succeeded in breaking off with England."

The chief, with his two gnarled hands in his lap and his head bowed, spoke of the great General. Tears had sprung to his eyes and he stopped talking and looked out the window.

Hardened, tough journalist lady that I am, tears came to my eyes, too. I had to fumble in my pocketbook and pull out a tissue to dab my eyes. Darn. I was with the chief less than ten minutes and I was already sniffling in my hanky.

The Indians always make me cry. Like blacks do when they sing about sweet Jesus or the songs they sang when they worked as slaves in the fields. Or when I look at pictures and into the eyes of early women pioneer feminists like Susan B. Anthony, and see that look staring out from the pages of history. Or see photos of my grandfather who worked in British factories from dawn to dusk at just nine years old. Oh, I cry all right. I cry for the past suffering, courage, and indomitable faith.

"My father, Chief George Forest Custalow, brought the first load of pulpwood to Elis Olsson when he bought the paper mill in West Point in 1917," the Chief said.

"Did the wood come from the Indian reservation?" I asked stupidly while busy taking notes.

"All the land was Indian land at one time, Sugar," he answered kindly. I blushed in acute embarrassment.

"I used to sit next to my father and we carried the wood down with a team of horses in an old wood wagon. I was just a boy back then. I can still see the old mill on the bank of the river. It looked like a coffee pot sitting in the marsh with its one great big smokestack standing up tall. My father was a close friend of old Elis Olsson, and the Mattaponi tribe was the main source of wood in those days."

The chief stopped to gather his thoughts. "When Elis Olsson came into West Point and took over the little paper mill and formed the Chesapeake Company, no one knew for sure if the Swede could get his new equipment to work. It took time before the new machine could get started. The first sticks of wood that went into Olsson's new paper pulp machine came from the woods behind the Mattaponi Indian Museum." At one time the chief's father was the sole wood agent for Mr. Olsson.

In 1925, the chief's father purchased a model T Ford truck and Webster drove the first load of a half cord of pine into the mill himself as a young teen-age boy. Changing over from horse and wagon to truck delivered wood was a quantum leap for the early wood dealers.

In those days Webster and his father and brothers chopped down every single tree themselves with an old fashioned ax and cut the wood into eight foot lengths, later changing to five-foot lengths better suited to trucks. Later they used a one-man "buck" saw and later the two-man saw came in. The new saws were a big improvement over the old ax cuts because they left the tree in better condition.

They were strong men and worked hard every day of the year in order to meet the needs of the new mill. He and his brother could saw a log and heave it up on the truck by themselves. They worked long and hard hours every day of the year bringing in the wood.

The chief knew Elis Olsson appreciated everything the Indians did for the Company. One day in about 1930, the state police stopped Custalow's truck just outside the front gate. "Olsson came flying out of his office like a wild horse, telling that policeman not to bother anyone bringing in wood to his Company and that this was Chesapeake property and the police must never stop any of his men," the chief said. When the law left, Olsson told the chief, "If the police ever bother you again for any reason, you call on me personally."

Chief Custalow was born in 1912, grew up on the reservation of the oldest tribe in the country (dating back to 1652) and was graduated from King William High School. After high school he felt fortunate to get a job at the local pickle plant in West Point, where he worked for five cents an hour, or 50 cents a day. One day the bossman told him he was such a good worker he would put him on salary for $6 a week. Trouble was he discovered the "weeks" were really two weeks long!

He saved his money and bought a few trucks and started working for the highway department on the new roads in the area. Because he owned several trucks, he was able to become a wood agent. Cecil Woodward of Chesapeake Corporation hired him in 1932 and there he stayed for the rest of his working life.

"Woodward told me I hauled in more wood for Chesapeake than any other agent," the Chief said. "I dedicated my life to making sure the mill never ran out of wood."

The chief and his men were used to a tradition of hard work. "We kept the wood coming into the mill even when the snow was deep on the ground and the river iced up so thick the boats could not get through to pick up the wood."

Chief Custalow said it was his personal duty to see that the mill never closed and, over the years, many crews worked for him and helped him meet his commitments. Sometimes his crews worked on and on, even when no one else could get a crew out on the road.

One time, the chief remembered, Mr. Woodward called him up and the weather was so bad he said, "Chief, if you can get men out working in the woods today, tell them I will personally give each of them a pair of new boots, compliments of Chesapeake Corporation. Then Woodward called up George Ashley in West Point, who sold boots, and bought every last pair of boots in West Point that day."

The chief smiled at the memories. "But my men came out and they always came out and they worked hard. The wood measurer came out to measure wood at 7 sharp every morning in front of the mill and my men were there at the head of the line every morning at 6 sharp. That was the way we did things, always worked hard, always did what we said we would do, all for to keep the Company going because without the mill West Point would have had no jobs and no money."

The chief thought a few moments and then continued with his tale. "I remember when a cord of wood only brought in 35 cents and men would chop wood for five cents an hour. When I first went to Chesapeake, Woodward paid me $4.25 a cord and I would pay 75 cents to the landowner for a cord of wood. When I had to hire people, Chesapeake helped me pay social security. Those were the days. Of course you could fill up a model T truck for a dollar for gas and get a dime change, too. Three dollars would fill up the truck with a whole load of groceries."

The chief told how they developed a good system on the river to get the wood down to the mill by building "chutes" on the high bank where they could take the wood to the landing and just push it down the chute. The boats would take it into the mill.

There were many days when the weather was cold for working outside. "The winter of '36 was so bad the river froze up and the ice was 25 inches thick and the boats couldn't move. So we got out our trucks and we worked day and night and the motors never cooled, we worked so hard," the chief remembered.

During World War II, the chief supervised German prisoners at Sandy Point Landing and the Germans were good workers. "They worked hard and all they wanted after a hard day was a pack of American cigarettes.

"Mann Bland, Chesapeake's wood measurer, would come over to the landing every week and measure all the cut wood standing in the racks and I remember how he would mark the ends of the measured wood with a piece of red chalk to make sure he wouldn't measure them again and then he would figure up how much he owed us and pay us right on the spot."

In those days, the Company had small boats that could get up in shallow water upriver as far as Aylett. Two small boats that served many years of service were the "Ethel" and the "Vamp," captained by George Galasky and John Smith.

The chief saw many changes in his long years of service. Eventually the trucks and hydraulic equipment were improved so much that nowadays loggers can go into the woods and cut, strip, and load trees without human hands ever touching a piece of wood.

"By 1972, my arthritis and other health problems caused me to retire from the wood hauling business," the chief said. "In all those years I never saw more of an injury in the woods than a chopped off toe. But I saw many close calls in the woods when those big trees were coming down."

The time had come for my interview to come to an end. The old chief had told me quite a story about the early days in this area. I stood up to say good-bye to Chief Custalow. We shook hands again. Then I instantly realized how the old Chief knew I was not local. Virginia women don't shake hands with men as equals the way Yankee women do.

"Good-bye, Sugar. Come again," the old chief called.

Before I pulled out of the drive, I added one conclusion to my copious notes. "Like General George Washington, Chesapeake Corporation could not have made it through those early years without the help of the Mattaponi Indians."

(Reprinted with permission from the Southside Sentinel, Urbanna, Va., 1996)

About the Author:

Ohio native Mary Wakefield Buxton is the author of seven previous books about life in Virginia. A graduate of the College of William and Mary, she also holds a M.Ed. from George Washington University with a major in human resource development. She and her husband, Chip, live on the Rappahannock River in Urbanna, Virginia, where she writes a column for the weekly Southside Sentinel titled "One Woman's Opinion."

About the Cover:

The photograph on the cover is of an original watercolor painting by Casey Holtzinger, picturing the tugboat "Sture" and Chesapeake barge Number 14. The painting, which was altered to fit the cover, is owned by Paul Harper of West Point, Virginia.